The Nervous System and Sense Organs

Julie McDowell

HUMAN BODY SYSTEMS
Michael Windelspecht, Series Editor

Greenwood Press
Westport, Connecticut • London

Library of Congress Cataloging-in-Publication Data

McDowell, Julie.
   The nervous system and sense organs / Julie McDowell.
      p. cm.—(Human body systems)
   Includes bibliographical references and index.
   ISBN 0–313–32456–5 (alk. paper)
   1. Nervous system.   2. Nervous system diseases.   3. Sensation—physiology.   I. Title.
II. Human body systems.
  QP355.2.M39   2004
  612.8—dc22   2003067638

British Library Cataloguing in Publication Data is available.

Library of Congress Catalog Card Number: 2003067638
ISBN: 0–313–32456–5

First published in 2004

Greenwood Press, 88 Post Road West, Westport, CT 06881
An imprint of Greenwood Publishing Group, Inc.
www.greenwood.com

Printed in the United States of America

The paper used in this book complies with the
Permanent Paper Standard issued by the National
Information Standards Organization (Z39.48–1984).

10 9 8 7 6 5 4 3 2

Illustrations, unless otherwise credited, are by Sandy Windelspecht.

The *Human Body Systems* series is a reference, not a medical or diagnostic manual. No portion of this series is intended to supplement or substitute medical attention and advice. Readers are advised to consult a physician before making decisions related to their diagnosis or treatment.

Contents

Series Foreword   ix

Acknowledgments   xiii

Introduction   xv

Interesting Facts   xvii

1 • **Nerve Cells: The Foundation of the Nervous System**   1

2 • **The Spinal Cord**   19

3 • **The Brain**   27

4 • **Peripheral and Autonomic Nervous System**   45

5 • **The Senses**   59

6 • **History of the Discovery of the Nervous System**   75

7 • **Nobel Prize Winners in Medicine and Physiology
Pursue Neuroscience Discoveries**   89

8 • **Nervous System Diseases and Disorders** 103

9 • **Keeping the Nervous System Healthy** 125

10 • **Future Avenues of Researching the Nervous System** 151

Acronyms 169

Glossary 171

Organizations and Web Sites 189

Bibliography 193

Index 197

Color photos follow p. 102.

Series Foreword

Human Body Systems is a ten-volume series that explores the physiology, history, and diseases of the major organ systems of humans. An organ system is defined as a group of organs that physiologically function together to conduct an activity for the body. In this series we identify ten major functions. These are listed in Table F.1, along with the name of the organ system responsible for the activity. It is sometimes difficult to specifically define an organ system, because many of our organs have dual functions. For example, the liver interacts with both circulatory and digestive systems, the hypothalamus acts as a junction between the nervous and endocrine systems, and the pancreas has both digestive and endocrine secretions. This complex interaction of organs and tissues in the human body is still not completely understood.

This series is unique in that it provides a one-stop reference source for anyone with an interest in the human body. Whereas other references frequently cover one aspect of human biology, from anatomy and physiology to the prevention of diseases, this series takes a more holistic approach. Each volume not only includes a physiological description of how the system works from the cellular level upward, but also a historical summary of how research on the system has changed since the time of the ancients. This is an important aspect of the series, and one that is frequently overlooked in modern textbooks. In order to understand the successes and problems of modern medicine, it is first important to recognize not only the achievements of the past but also the misunderstandings and challenges of the pioneers in medical research.

For example, a visit to any major educational institution reveals large lec-

TABLE F.1. Organ Systems of the Human Body

Organ System	General Function	Examples
Circulatory	Movement of chemicals through the body	Heart
Digestive	Supply of nutrients to the body	Stomach, small intestine
Endocrine	Maintenance of internal environmental conditions	Thyroid
Lymphatic	Immune system, transport, return of fluids	Spleen
Muscular	Movement	Cardiac muscle, skeletal muscle
Nervous	Processing of incoming stimuli and coordination of activity	Brain, spinal cord
Reproductive	Production of offspring	Testes, ovaries
Respiratory	Gas exchange	Lungs
Skeletal	Support, storage of nutrients	Bones, ligaments
Urinary	Removal of waste products	Bladder, kidneys

ture halls, where science instructors present material to the students on the anatomy and physiology of the human body. Sometimes these classes include laboratory sessions, but in the study of human biology, especially for students who are not bound for professional schools in medicine, the student's exposure to human biology typically centers on a two-dimensional graphic. Most educators accept this process as a necessary evil of the educational system, but few recognize that, in fact, the large lecture classroom is the product of a change in Egyptian religious beliefs before the start of the current era. During the decline of the Egyptian empire and the simultaneous rise of the ancient Greek culture, the Egyptian religious organizations began to forbid the dissection of the human body. This had a twofold influence on medicine. First, the ending of human dissections meant that medical professionals required lectures from educators instead of participation in laboratory-based education, which led to the birth of the lecture hall. The second consequence would plague modern medicine for a thousand years. Stripped of their access to human cadavers, researchers studied other "lesser" animals and extrapolated their findings to humans. The practices of the ancient Greeks were passed on over the ages and became the basis for the study of modern medicine. These traditions continue to this day throughout the educational institutions of the world.

The history of human biology parallels the development of modern science. In the seventeenth century, William Harvey's study of blood circulation challenged the long-standing belief of the ancient Greeks that blood was produced in the liver and consumed in the tissues of the body. Harvey's pioneering experimental work had a strong influence on others, and within a century the legacy of the ancient Greeks had collapsed. In the eighteenth century a group of chemists who focused on the chemical reactions of the human body, called the iatrochemists, began to apply chemical laws to human physiology. They were joined by the iatrophysicists, who believed that the human body must operate under the physical laws of the universe. This in turn led to the beginnings of organic chemistry and biochemistry in the nineteenth century, as scientists focused on identifying the building blocks of living cells and the chemical reactions that they utilize in their metabolism.

In the past century, especially in the last three decades, the rapid advances in technology and scientific discovery have tended to separate most sciences from the general public. Yet despite an ongoing trend to leave the majority of the physical sciences to the scientists, interest in the human biology has actually increased among the general population. This is primarily due to medical discoveries that increase not only lifespan but also healthspan, or the number of years that people live disease free. But another important aspect of this trend is the desire among the general public to be able to ask intelligent questions of their physicians and seek additional information on prescribed medications or procedures. In many cases this information serves as a system of checks and balances on the medical profession, ensuring that the patient is kept well informed and aware of the fundamentals regarding the procedure.

This is one of the most remarkable ages in the study of human biology. The recent announced completion of the Human Genome Project is an indication of how far biology has progressed. Barely fifty years ago, scientists were first discovering the structure of DNA. They now are in possession of an entire encyclopedia of human genetic information, and although they are not yet exactly sure what the content reveals, scarcely a week goes by without a researcher announcing a medical discovery that was made possible by the availability of the complete human genetic sequence. Coupled to this are the advances in the development of pharmaceuticals and treatments that were unheard of less than a decade ago.

But these benefits to society do not come without a cost. The terms stem cells, cloning, and gene therapy no longer belong to the realm of science fiction. They represent advances in the sciences that may hold the key to increased longevity. However, in many cases they also produce ethical and moral questions of society: Where do medical researchers obtain the embryonic stem cells for their work? Who will determine if humans can be cloned? What are the risks of transgenic organisms produced by gene therapy? These are just a few of the potential conflicts that face modern society. Only a well-educated general public can intelligently survey the pros and cons of an ethical or moral decision regarding medical science. Armed with information, concerned people can participate in the democratic process of informing their elected officials of their concerns. Science education is an important aspect of citizenship, and thus the need for series such as this to present information to the general public.

This volume covers the biology of the nervous system, an amazing information processing center that controls the majority of people's everyday lives. Although people may be well aware of sensory input such as light, heat, and sound, in reality the nervous system is being bombarded by millions of inputs per second. Atmospheric pressure, internal temperature, muscle position and load, and the acid levels of the body fluids are just a few of many inputs constantly being monitored. The majority of the nervous system's work is both involuntary and automatic, meaning that people do not need to devote limited conscious thought to the processing, or integration, of sensory input. The nervous system also represents an area of intense medical research. Nervous system disorders such as Alzheimer's disease and Parkinson's disease rob the elderly of their memories and ability to move late in life. Physicians and researchers are just beginning to explore the genetic basis of nervous system disorders, and already have made much progress in the development of pharmaceutical drugs to treat the symptoms of many nervous system diseases. Given the prevalence of these diseases in the general population, there is a definite need for a volume such as this that explores the history, physiology, and diseases of the nervous system.

The ten volumes of *Human Body Systems* are written by professional authors who specialize in the presentation of complex scientific ideas to the general public. Although any book on the human body must include the terminology and jargons of the profession, the authors of this series keep it to a minimum and strive to explain the concepts clearly and concisely. The series is ideal for the public libraries, as well as for secondary school and introductory college libraries. In addition, medical professionals or anyone with an interest in human biology would find this series a useful addition to their personal library.

Michael Windelspecht
Blowing Rock, North Carolina

Acknowledgments

Writing, researching, and producing this volume of the *Human Body Systems* series has been both an invigorating challenge as well as a gratifying, and at times, exhausting experience. I am indebted to the editor of this series, Michael Windelspecht, for his invaluable support, assistance, and encouragement throughout the entire process. His feedback and guidance not only helped me with this project, but has helped me develop as a professional science writer. I am also deeply thankful to this volume's illustrator, Sandy Windelspecht, who was able to make art from my rough ideas and managed my repeated revisions and change suggestions. Thanks also is due to Elizabeth Kinkaid for gathering the photos and supplemental graphics. In addition, I am grateful for all the support and encouragement from Debby Adams at Greenwood, who was wonderful at dealing with my barrage of e-mails and questions.

Finally, I would like to thank my parents, Douglas and Sharon McDowell, and my sister, Christine, for their endless supply of love, friendship, and support.

Introduction

Learning and understanding the human body's nervous system can be daunting. It is no wonder that some of history's greatest minds struggled their entire lives to solve the mysteries of how the brain operates and how the body reacts to its surroundings. The great Greek intellectuals—including Aristotle, Democritus, and Hippocrates—argued back and forth about the workings of the universe and how the human body fit in and functioned.

Although the human body often mystified these scientific detectives, they were all driven by a need to know how the body feels, thinks, and moves. They wanted to know how they were able to hear music, taste the flavor in food, and smell the aroma of flowers. They wanted to know how their bodies knew to sweat when it was hot and how their hands knew to pull away when they touched a sharp object (such as a knife). These are the questions that drove scientists—from ancient Greece to our contemporary society—to discover the workings of the central nervous system and how the brain and spinal cord affected the quality of life.

After scientists determined how the nervous system processed information and allowed the body to function, research also became focused on how to heal and treat the body when it was in the throes of disease, or after an injury. But the understanding and advancements in medicine and treatment for nervous-system-related disorders are inextricably linked. The more that is understood about the body's nervous system, the more that can be understood about how it can break down. Research on diseases and disorders of the nervous system allows scientists to learn why the illness is present, how medication can be best targeted for treatment, and, ideally, how the treatment can lead to a cure. For instance, the more scientists learn about spinal cord injuries and paralysis, the closer they get to finding a cure and helping the disabled to walk again.

In order to fully understand how everything in the body functions, it is necessary to learn its basic foundations. An understanding of all of the nervous system's fundamental elements will make it easier to tackle the more advanced information, such as nervous-system-related diseases. Following a list of interesting facts about the nervous system and sense organs, the first chapter of this volume introduces the cellular basis of the nervous system, which includes neurons and nerve systems, in addition to how an impulse is received by the neurons and then translated into a reaction. Chapter 2 breaks down the workings of the spinal cord, and how this major organ works with the brain to process information and impulses. Next, the third chapter focuses on the various parts of the brain, and how they work together to complete the nervous system. The fourth chapter is devoted to the peripheral and autonomic divisions of the nervous system, which include the brain-related and spinal nerves, in addition to the body's automatic, or involuntary, reactions and functions. Chapter 5 covers the senses—how the nervous system regulates how we smell, taste, hear, see, and feel.

Chapters 6 and 7 look at the history of the nervous system's discovery, from early researchers, such as Leonardo da Vinci, to contemporary scientists who were recognized with a Nobel Prize for their nervous system research, or neuroscience work. Chapter 8 is devoted to some of the diseases and disorders of the nervous system, and Chapter 9 looks at some important elements—including nutrition and exercise—that keep the nervous system healthy. Finally, Chapter 10 looks at how research is looking to treat and possibly cure nervous-system-related diseases and disorders. Each of these chapters features tables, pictures, and illustrations that are meant to enhance your learning experience.

At the end of the volume is a list of acronyms used in the book, followed by a glossary that highlights important concepts and terms, a list of organizations and Web sites, a bibliography, and an index.

This volume is designed to foster an understanding of one of the body's key operations, in addition to related information in order to provide a comprehensive look at an important element of the human body. The target audience is the general public and secondary and undergraduate students. With this audience in mind, an attempt has been made to use common language to describe medical and scientific concepts wherever possible. The glossary in the back of the volume provides definitions for key terms, which are indicated by **bold** type on their first use in the text.

INTERESTING FACTS

▶ Average weight of an adult human brain: 2.8–3.1 pounds (1,300–1,400 grams)

▶ Average weight of an elephant's brain: 17.2 pounds (7,800 grams)

▶ Average number of neurons in the brain: 100 billion

▶ Length of myelinated nerve fibers in the brain: 93,200–112,000 miles (150,000–180,000 kilometers)

▶ Difference in the number of neurons in the brain's left and right hemispheres: 186 million more neurons in the left hemisphere in comparison to the right hemisphere

▶ Total surface area of the human brain's cerebral cortex: 2.5 square feet (2,500 square centimeters)

▶ Total surface area of an elephant's cerebral cortex: 6.8 square feet (6,300 square centimeters)

▶ Total number of neurons in the cerebral cortex: 10 billion

▶ Total volume of cerebrospinal fluid in the human body: 4.2–5.1 fluid ounces (125–150 milliliters)

▶ Half life of cerebrospinal fluid: 3 hours

▶ Daily production of cerebrospinal fluid: 13.5–16.9 fluid ounces (400–500 milliliters)

▶ Number of neurons in the human spinal cord: 1 billion

▶ Length of human spinal cord: male—17.7 inches (45 centimeters), female—16.9 inches (43 centimeters)

▶ Weight of human spinal cord: 1.2 ounces (35 grams)

▶ Total number of human taste buds on the tongue, palate, and cheeks: 10,000

▶ Number of taste buds on the tongue: 9,000

▶ The skin contains 45 miles (72.4 kilometers) of nerves

▶ Nerve impulses can travel from the brain at speeds up to 170 miles (274 kilometers) per hour

Nerve Cells: The Foundation of the Nervous System

The human body's nervous system is divided into two parts: the **central nervous system** (CNS) and the **peripheral nervous system** (PNS). The CNS consists of the brain and the spinal cord, and the PNS consists of the **cranial nerves** (the brain's twelve pairs of nerves) and the **spinal nerves** (thirty-one pairs of nerves associated with the spinal cord). Also included in the PNS is the **autonomic nervous system** (ANS), which controls the "automatic" or involuntary movements of the body's smooth muscles (found in the walls of tubes and hollow organs), cardiac muscles, and glands (see Figures 1.1 and 1.2 for a general overview of the organization of the nervous system and the somatic nervous system that controls voluntary movement). The two divisions of the ANS are the **parasympathetic division**, which dominates and controls the body during nonstressful situations, and the **sympathetic division**, which dominates and controls the body during stressful situations.

Through the study of evolutionary trends, researchers determined that over 400 million years ago (during the Ordovician period of the Paleozoic Era), the simplest **vertebrates** had a CNS composed of a neural tube that ran the length of the body, an early version of the spinal cord now present in the human body. This neural tube was later surrounded by a bony encasement known as a spinal or vertebral column, which is made up of separate bones called the **vertebrae**. Over geological time, the front end of the neural tube expanded into three swellings—this became the brain.

Information is sent from the PNS to the brain, which serves as the activ-

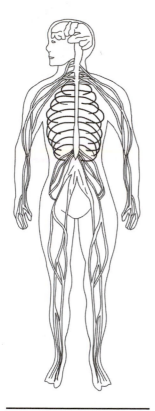

Figure 1.1. The nervous system.
The brain and nerve systems, including the spinal cord, make up the nervous system. While regulating movements, the nervous system also works to interpret sensory information.

ity headquarters of the CNS. Through the five senses (sight, smell, touch, hearing, and taste), the CNS detects a **stimulus**, which is a change that prompts a response in a living organism. The brain then processes the transmitted information and initiates the appropriate response or responses in the effector. An **effector** is an organ, such as a muscle or gland, that responds through some kind of reaction (usually movement) when receiving a stimulus.

In the mid-nineteenth century, scientists began studying the human body's nervous system by using a microscope. The invention of the microscope is widely attributed to Zacharias Janssen (1580–1638), a Dutch spectacle maker who introduced the concept of using a combination of more than one lens for magnification during the late sixteenth century. But Janssen's model only provided ten times magnification. Shortly after Janssen's invention, a Dutchman by the name of Antony van Leeuwenhoek (1632–1723) and a British scientist by the name of Robert Hooke (1635–1703) independently refined these compound microscopes, which are microscopes that use more than one lens. With this ability for greater magnification, Hooke and van Leeuwenhoek published their observations and illustrations on structures of microorganisms, including their cellular construction. Following these discoveries, researchers were able to determine two kinds of cells in the nervous system: neurons, and other cells that are responsible for taking care of the neurons—including feeding and repairing them (see Chapter 6 for a more detailed history of the nervous system research and discoveries).

A **nerve** is defined as a group of nerve fibers located outside the CNS, and bundles of nerve fibers located within the CNS are called **nerve tracts**. Nerve fibers make up nerve cells, also known as **neurons**, which are the building blocks of the entire nervous system. Neurons are the essential conducting unit of the entire nervous system.

Before looking at the details of the neuron, it's important to understand some other important elements that enable the CNS to operate smoothly. Taking a close look at a slice of the spinal cord (Figure 1.3), one sees that most of it is composed of white material and some of it is composed of a gray substance. These two parts are simply called **white matter** and **gray matter**. Whereas the gray matter consists mostly of nerve cell bodies, the white matter is composed mostly of nerve fibers.

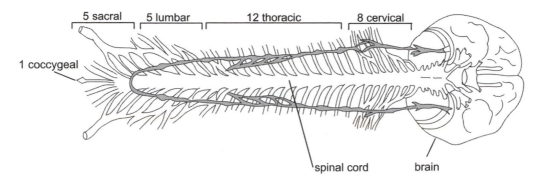

Figure 1.2. The spinal cord.
There are five categories of spinal nerves noted on top of the figure: the coccygeal, sacral, lumbar, thoracic, and the cervical, which total thirty-one pairs of nerves.

NERVE CELLS

During the initial stages of growth, nerves develop in the embryo's CNS and then grow out and spread through the body like tentacles. There are thousands of nerve fibers grouped in large bundles that run to and from the CNS. **Afferent nerves** are those fibers coming to the CNS from the muscles, joints, skin, or internal organs, whereas those leaving the CNS to travel to these areas are known as **efferent nerves**. The spinal cord has two types of afferent nerves: those coming in at the back are called posterior or **sensory**; those leaving the spinal cord come out the front and are called anterior or **motor**. When these afferent nerves reach the spinal cord's white matter, they divide and branch out to bring their messages to various different areas of the cord. Most of the branches connect with neurons near the entry region of the nerve fiber and intermediate neurons. These intermediate neurons then connect with the motor neurons, which control muscle movement.

Although neurons vary in size, shape, and functions, they all consist of the following four distinct parts:

1. *Cell body.* This is the main mass of the cell and contains the nucleus and other **organelles**. The **nucleus** is the cell's largest organelle, which contains **chromosomes**. In these chromosomes, genes carry the body's hereditary information in the **DNA**, which allows the cell to reproduce. Neuron cell bodies are found in the CNS or close to it in the trunk of the body so they are protected by bone. The **cell body** also contains **cytoplasm**, and an abundance of **mitochondria**, which are organelles responsible for energy production in the cell.

2. *Dendrites.* This group of branching nerve fibers carries impulses to the cell body. Small black spots also appear on the **dendrites**: these represent the nerve endings of other neurons, which pass messages from other neurons.

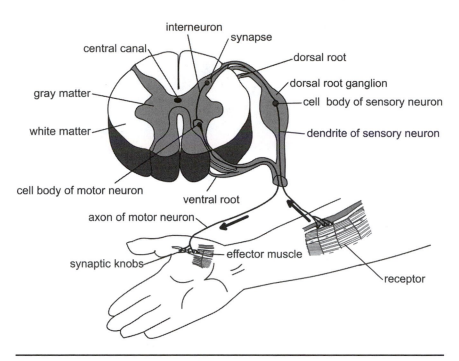

Figure 1.3. A cross-section of the spinal cord.
A cross-section of the spinal cord displays the white matter, which is the nerve tissue that contains myelinated axons and interneurons, and the gray matter, which is the nerve tissue that contains the neurons' cell bodies. The spinal roots—the dorsal and ventral roots—are just shown on the right side of the spinal section, although these roots and the corresponding nerve tracts are actually present on both sides of the spinal cord. This figure also shows the path of a reflex arch, with the receptor muscles receiving the impulse, which is processed by the sensory neuron and then transmitted to the motor neuron at the synapse in the spinal cord.

3. *Axon.* This single nerve fiber carries impulses away from the cell body and the dendrites. In humans, some **axons** can be around 1 meter long, but in larger mammals such as whales and giraffes these would be much longer. A **nerve fiber** is the neuron including the axon and the surrounding cells. These fibers branch out at the neuron's ending (also known as arborization) and can be classified as either **excitatory** or **inhibitory**.

4. *Transmitting region.* The axon carries the impulses to the transmitting region, and from here it leaves the cell body and travels to the CNS.

In addition to these four parts, neurons are composed of layers of membranes and microtubules that produce hormones, neurotransmitters, and substances such as **peptides** and proteins, which will be discussed in detail further in this chapter.

HOW CELLS PRODUCE ENERGY

Cells produce energy through **aerobic cell respiration**. The following equation provides a simple explanation:

$$\text{Glucose } (C_6H_{12}O_6) + 6O_2 \rightarrow 6CO_2 + 6H_2O + \text{ATP} + \text{heat}$$

Each product of cell respiration is vital to the body's function. The heat regulates body temperature, the water feeds the cells, and the CO_2 is the waste that's exhaled through breathing.

An important product of cell respiration is **ATP (adenosine triphosphate)**, a **neurotransmitter** that is the muscle's direct source of energy for movement. Basically, ATP captures energy from food, breaks it down, and then releases it into cells. Some of this energy resulting from respiration is used by the cell's mitochondria to produce ATP. Therefore, cellular respiration is a constant, continuing cycle. All cells contain molecules of **ADP (adenosine diphosphate)** and phosphate. When the body digests food, it breaks down various chemical components to use in cell processes. A form of sugar known as **glucose** $(C_6H_{12}O_6)$ is one such substance, and is a necessary component (along with oxygen) in aerobic respiration. Glucose breaks down into CO_2 and H_2O (water), along with a release of energy. This energy then bonds with the ADP and a third phosphate to form ATP. Energy for cell processes is released and available for use when the bond holding this third phosphate is once again broken down.

NEURONS

As mentioned earlier, neurons are nerve cells, that are composed of nerve fibers. Neurons are the foundation of the entire nervous system, and are responsible for transmitting messages throughout the body. In the PNS, the neurons are constantly carrying information to and from the CNS. However, neurons can only carry electrical impulses in one direction, making it impossible for impulses to run into each other and cancel each other out.

The nervous system is made up of millions of neurons, in addition to the other cells that help support the functions of the spinal cord and brain. Neurons are classified into three groups: sensory, motor, and interneurons. Sensory and motor neurons make up the PNS, and interneurons are found in the CNS.

The **sensory** or **afferent neuron** mainly functions in the CNS, and the **motor** or **efferent neuron** mainly functions in the PNS. Although both these neurons contain the same physical parts (the cell body, dendrites, axon, and trans-

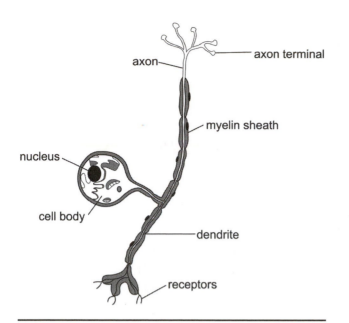

axon terminal

axon

myelin sheath

nucleus

cell body

dendrite

receptors

Figure 1.4. A sensory neuron.

This neuron structure details a sensory neuron, which carries impulses and messages to the spinal cord and brain. These sensory neurons, which have cell bodies, are located in the CNS and close to the body so they are protected from damage. The axon moves these messages away from the nucleus and cell body, and the dendrites work in reverse to bring the impulses into the cell body. The myelin sheath works as an insulator to protect the neurons from short-circuiting with each other.

mitting region), there are some important differences in their composition (Figures 1.4 and 1.5).

The sensory or afferent neuron carries impulses and messages to the spinal cord and brain. In these neurons, the dendrites act as receptors—they detect external or internal changes, and transmit this information to the CNS. As evident in Figure 1.4, the sensory neuron dendrites may be short or long (sometimes as long as 1 meter), but they are single and are not branchlike in appearance like in the motor neuron.

Once the information is transmitted to the brain and spinal cord, the impulse is interpreted and the CNS stimulates a sensation. Sensory neurons located in the skin, skeletal muscle, and joints are known as **somatic**, and those in the internal organs are called **visceral**.

The motor (or efferent) neuron (Figure 1.5) carries the impulses from the CNS to muscles and glands, also known as effectors. After the CNS processes an impulse, the motor neuron will tell muscles to contract or relax and glands to secrete. Those motor neurons associated with the skeletal muscle are called somatic, and those associated with the smooth muscle, cardiac muscle, and glands are called visceral.

Some axons in both in both the CNS and PNS are layered with a covering called the **myelin sheath**. Composed of fatty material, the myelin sheath electronically insulates neurons from one another. Without the protection of the myelin sheath, the neurons would short circuit and thus be unable to transmit electronic impulses.

In the motor neurons, the axons and dendrites are wrapped in specialized cells called **Schwann cells**, which create their myelin sheath. The nucleus and cytoplasm in the Schwann cells are collectively called the **neurolemma** and are located outside the myelin sheath, physically covering the nerve cell. Schwann cells are located in the PNS. The specialized cells in the CNS are called **neuroglia**. The **node of Ranvier** is the space that separates the

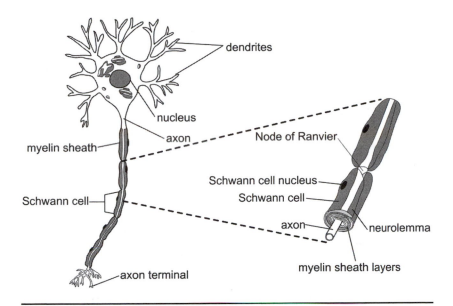

Figure 1.5. A motor neuron.
This neuron structure details a motor neuron, which carries the impulses from the central nervous system to muscles and glands. Compared to the sensory neuron, this structure features the cell body at the top (rather than the side) and does contain receptors.

Schwann cells. These nodes are responsible for depolarizing electrical impulses, which will be explained in greater detail later in this chapter.

The neurolemma is important for nerve regeneration. If a nerve or nerves in the PNS is damaged or severed (for example, if a limb is damaged or severed) and then reattached through surgery, the neurolemma allows the axons and dendrites to regenerate and reattach to their proper connections. In addition, the Schwann cells are believed to produce a chemical substance that stimulates regeneration. This regeneration may be a slow process (it could take months or years), but eventually the nerve fibers may restore their functions, therefore reinstating feeling and movement to the limb.

Regeneration, however, is not possible in sensory neurons of the spinal cord because there are no Schwann cells. In these neurons, the myelin sheath is formed by the **oligodendrocytes**, the specialized cells found only in the brain and spinal cord. Because there are no Schwann cells, there is no neurolemma, and thus cell regeneration is impossible. This is why spinal cord damage or severing results in permanent loss of function (see Chapter 8).

The final classification of neurons is called **interneurons**. These are located entirely within the CNS and combine or integrate the sensory and motor impulses. Some of their functions include thinking, memory, and learning. For instance, interneurons might receive impulses from the brain and transmit them to the somatic nervous system that determines movement in voluntary muscles, such as fingers and toes.

Cranial nerves, or the nerves located in the brain, contain sensory, motor, and mixed nerve fibers. The sensory nerves carry impulses toward the brain, and the motor nerves carry impulses away from the brain. But most of the cranial nerves and all of the spinal nerves are made up of both sensory and motor fibers, and these are called **mixed nerves**.

SYNAPSES

Every neuron has dendrites that connect with other neurons. In the CNS, neurons communicate with each other when the axon of one neuron comes into contact with the cell body or dendrite of another neuron. The space or junction between the axon and dendrite of these two neurons is known as a **synapse**, which comes from the Greek word meaning "to clasp." This is where the message carried by the neuron is passed on. The synapse is often called a relay because it is here where the information is relayed to the next neuron. More specifically, the actual area (which is between 10–50 nm in width) between the axon and dendrite is known as the **synaptic gap** or **cleft**.

Like a blueprint showing the floorplan and layout of a house, the location and pattern of connections between neurons determine the structure and organization of the CNS. The position of the synapses dictate the route that impulses will follow within and between the brain and spinal cord. The impulse pathways determine what sensations are experienced and how an effector responds to these sensations.

Events related to the synaptic process are classified as either **presynaptic** (before transmission) or **postsynaptic** (following transmission). It's important to note that neurons conduct impulses in only one direction, depending on if they are afferent or efferent.

The synapse transmissions in the CNS are extremely complicated, but Figure 1.6 describes their role in transporting information throughout the CNS at the most basic level. In this illustration are four neurons—A, B, C, and D. A and B are the presynaptic neurons that are carrying information to C and D, the postsynaptic neurons. The A and B axons will make contact with the C and D dendrites, thus creating four synapses.

Neurons connect with each other in two ways: (1) a neuron receives impulses from a few other neurons and relays these impulses on to thousands of other neurons—**divergence**; and (2) a neuron receives impulses from the nerve endings of thousands of other neurons and transmits its message to only a few other neurons—**convergence**. Convergence and divergence are two types of pathways that impulses can take when traveling from the presynaptic neuron to the postsynaptic neuron, as shown in Figure 1.6. An impulse can travel across synapse 1 to C, or it can go across synapse 2 to D.

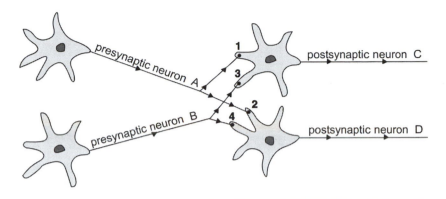

Figure 1.6. Synaptic connections.
This diagram describes the synapse transmission between the neurons of the CNS. The two presynaptic neurons, A and B, carry impulses and messages to the postsynaptic neurons, C and D. The locations where A and B make contact with C and D are known as synapses, which are numbered 1–4.

However, the impulse can also travel along A and cross synapses 1 and 2 to C and D, which is divergence. Convergence is when the impulse travels along to presynaptic neurons to one postsynaptic neuron; for example, if A and B impulses would travel through synapse 1 and converge to postsynaptic neuron C. Of course, in the complex body, any postsynaptic neuron can expect to have hundreds of axons from presynaptic neurons attach to its dendrites.

A nerve impulse enters the cell body by first attaching itself to that cell's dendrites. This then alters the neuron's excitability. The excitatory nerve fiber amplifies the energy from the impulse, but the inhibitory nerve fiber reduces this energy. Although the arrival of the impulse alters the excitability of the neuron, it doesn't necessarily result in a postsynaptic neuron reaction or firing. A reaction occurs only when the nerve fiber's axon has reached its **threshold level**. The value of this threshold varies with each nerve fiber and depends on the composition of the cellular fluid and the number of impulses recently received and conducted. For example, when one touches a warm stove with his or her fingers, the nerve impulse translating the corresponding sensation of warmth attaches itself to the dendrites of the neurons located in the fingertips, thus stimulating the neuron's excitability. The excitatory and inhibitory nerve fibers process this sensation through amplifying but also inhibiting the energy from the impulse. When the sensation (or warmth) increases beyond the threshold level of the nerve fiber's axon, a reaction occurs. So if the heat on the stove increases beyond a certain level, the nerve fibers will stimulate a chain of events that produce a reaction, such as pulling the fingers away from the stove.

Both excitatory and inhibitory neurons are managed by excitatory and inhibitory nerve fibers. Some postsynaptic neurons fire off most of the time, whereas others fire off less frequently. For example, neurons that work the muscles in the brain and spinal cord are in a constant state of excitability. Because the body's muscles must always be ready for action, there must be a constant flow of impulses between the brain and spinal cord. But neurons that control breathing are not constantly firing—they operate at a more rhythmic pace.

The respiratory and muscular responses are examples of postsynaptic excitation or inhibition because they take place immediately after an impulse transmission at the synaptic gap. Presynaptic excitation or inhibition occurs before the impulse reaches the synaptic gap. If an impulse is transmitted along an inhibitory nerve fiber, it will not deliver its full load of excitability at the synapse. In fact, this nerve fiber can completely block all nerve impulses from reaching the synapse.

Synapses can deliver messages either to thousands of neurons or only a few. The more synaptic exchanges there are in a transmission, the more opportunities there are for changes or modifications that can be made to that resulting reaction. For example, suppose there are two ways to get from Anytown, USA, to Nowhere, USA. The first route is a straight road with no turnoffs, but the second route has numerous crossings and connects with secondary roads that ultimately lead to Nowhere. Although the first route is quicker, there is no chance for changes or modifications. One can't change the route—it's a straight shot. But the second route, even though it might be slower, has numerous opportunities for change with all its turnoffs and back roads.

When a nerve impulse crosses the synaptic gap, each synapse contains substances to create adhesion between the presynaptic and postsynaptic membranes. When contacted by the impulse, the synaptic gap constricts as the width decreases and the concentration of the transmitter increases. Synapses that are not used often tend to cease functioning, but synapses that are used frequently tend to transmit impulses quickly.

NEUROTRANSMITTERS

In the latter part of the nineteenth century, there was a controversy over whether the vertebrates had a nervous system composed of a continuous network of neurons or of separate neurons (see Chapter 6). When it was eventually proven that the vertebrate's body is made up of separate neurons connected with synapses, the question remained about how the neurons related to each other.

Because nerve fibers are minute and only cover a small portion of the postsynaptic neuron's cell body, the amount of impulse and energy it can deliver is diminutive. This might be adequate for organisms such as fish or crustaceans—there is enough current delivered to inhibit or excite a post-synaptic neuron in these animals. However, in reptiles and mammals another mechanism is used to transmit impulses.

In the early 1900s, scientists and physiologists had concluded that the endings of nerve fibers emit chemical substances that influence the behavior of postsynaptic neurons.

Researchers noticed that by injecting certain **nerve tissue** with a substance called **adrenaline**, the sympathetic system (the motor nerve network that operates the blood vessels and sweat glands, along with some of the internal organs) was stimulated. They believed that the body was constantly tapped into a natural supply of adrenaline.

In the early 1900s, Otto Loewi (1873–1961) studied the chemicals associated with nerve impulses and the heart. It was common scientific knowledge at this time that stimulating a specific parasympathetic nerve could slow down and eventually cause the heart to stop beating. Loewi used a frog to simulate this scenario, and then he collected the fluid produced in the heart ventricle and injected it into the heart of another frog. When he injected this fluid into the heart of another frog, its heart rate also slowed. This substance was later found to be **acetylcholine**. In subsequent experiments, Loewi determined a substance that could also increase the heart rate. In 1948, this was discovered to be **noradrenalin**.

These landmark discoveries verified that chemical substances are emitted through nerve endings to help transmit messages. These chemicals are called neurotransmitters, or simply transmitters. In the human body, there are about 80 different neurotransmitters.

Neurotransmitters are classified into four groups:

> Amines: acetylcholine, noradrenalin, serotonin, dopamine
>
> **Amino acids**: glutamic acids, gamma aminobutyric acid (GABA)
>
> Purines: adenosine triphosphate (ATP)
>
> **Peptides**: enkephalins, dynorphin, endorphin

Some important neurotransmitters in our body include acetylcholine, **dopamine**, **norepinephrine**, and **serotonin**. Many neurotransmitters have a particularly wide distribution and vital roles. Noradrenalin, for example, transmits neurons from various regions in the brain, such as the cerebral **hemispheres**, the cerebellum, and the spinal cord. Noradrenalin increases the reaction excitability in the CNS and the sympathetic neurons in the spinal cord.

Another major transmitter is serotonin, which is an important distributor for the sensory **channels** in the CNS and in the expressions of emotion. Medications that alter mood or behavior, such as antidepressants, will be targeted at serotonin, and norepinephrine and will affect synapse transmission.

Dopamine is an important transmitter in the motor system, limbic system, and the hypothalamus. Parkinson's disease kills the cells that produce these transmitters, therefore impeding mobility and other motor functions. These transmitters are supplemented using a drug called L-dopa, which improves posture and motility, although it can't stop the tremors that accompany Parkinson's disease (see Chapter 8).

In simplistic terms, when a presynaptic neuron receives an electrical impulse, its axon releases a neurotransmitter, which then diffuses across the synapse and combines with the dendrites of the postsynaptic neuron. This generates an electrical impulse, which is then carried to the postsynaptic neuron's axon to the next synapse and so on. A chemical inactivator is located at the dendrite of this postsynaptic neuron to counteract the impulse generated by the neurotransmitter. Each neurotransmitter has a specific inactivator. For example, cholinesterase is the inactivator for acetylcholine, a neurotransmitter found in many of the body's muscular junctions (see Figure 1.7). The inactivator stops the continuous transmission of the impulse unless a new impulse is generated by a neurotransmitter at the first neuron. But let's look at this process more closely to further understand the transmission of a nerve impulse.

Neurons produce transmitters either in their nerve endings or their cell bodies. In the case of peptides, they are made in the cell bodies, then transported to the nerve endings where they are then converted into transmitters. Some unmyelinated nerves, such as those in the sympathetic system, emit transmitters along the course of the nerve rather than the nerve ending. Transmitters pass through little bulges called varicosities located along the nerve fiber. These varicosities move along the nerve fiber, similar to when a ball moves through a stocking.

In other cases, transmitters are stored in tiny **vesicles** of the nerve endings. When an impulse or action potential affects these endings, calcium (Ca^+) ions pass into the endings from the extracellular fluid from the neighboring neutron. This connects the vesicle with the membrane of the nerve ending, and then the transmitter connects with the synaptic cleft. The upper layer of this four-layered membrane is presynaptic membrane, and the synaptic cleft is between this layer and the second layer.

After the transmitter leaves the vesicle, it crosses the synaptic cleft and connects with the membrane or receptor site of the postsynaptic neuron. This receptor site is made up of various proteins, and those proteins that are specifically affected by a transmitter are known as receptors or targets of the transmitter. It's important to note that one transmitter can have different effects on different neurons. Acetylcholine can excite one neuron while inhibiting another. However, in vertebrates, GABA is always inhibitory and glutamate is always excitatory (see Figure 1.7).

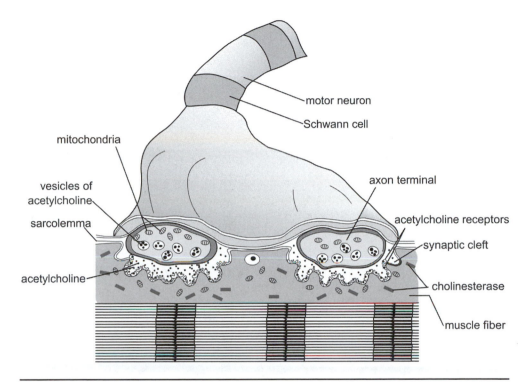

Figure 1.7. The neuromuscular function.

When the muscle fiber meets the motor neuron at the axon terminals, a movement or reaction results with the aid of neurotransmitters, including acetylcholine and its inactivator, cholinesterase. The sarcolemma contains receptors for the acetylcholine, and the mitochondria are where cell respiration takes place and energy is produced. Muscle contraction begins when the axon terminal receives a nerve impulse, which stimulates the release of acetylcholine. This release causes electrical changes—because of the movement of ions—at the sarcolemma.

Transmitters must be controlled. If their secretion was not stopped at some points, then rapid changes in excitation and inhibition would not occur and all activity would be slowed down. Some synapses block impulse transmission rather than pass the impulse on to other synapses (known as an **excitatory synapse**). A synapse that inhibits impulse transmission is known as an **inhibitory synapse**. In this situation, a chemical inactivator located at the dendrite of the postsynaptic neuron inactivates the neuron, which ceases any impulse unless a new impulse from the first neuron releases more of the neurotransmitter.

Peptides have been discovered to be transmitters only in the past twenty to thirty years. One example is the parasympathetic nerves that control the salivary glands, which release a transmitter called VIP, also know as vasointestinal peptide. VIP increases the amount of acetylcholine released by nerve endings and also dilates blood vessels, which brings more blood to the glands so more saliva can be secreted. In addition, because of its ability to dilate blood vessels, it's used in the penis to achieve an erection. Often someone who is impotent has a deficiency of VIP in his genital organs.

Another peptide transmitter is called substance P. This helps to excite neurons and nerve fibers to relay information gathered from noxious and thermal receptors to the brain, relaying messages that involve smells and heat.

NERVE IMPULSE

The body's nerve fibers system can be compared to a telegraph wire. Both are electric conduction systems designed to quickly relay messages over long distances. Both transmit messages in the form of pulses that are of a constant size and speed. Both the nerve fiber and the telegraph wire must be insulated for protection. If the wires or the fibers are damaged in any way, they can't carry the impulses. But the nerve fiber is significantly more complicated than a telegraph wire. The nerve fiber generates the message itself, along with transmitting it, whereas the telegraph wire is simply involved in the transmitting.

The body is composed of two kinds of fluid: **intracellular (ICF)** and **extracellular (ECF)**. ICF, the fluid within the cells, contains 65 percent of the body's total water. ECF, the fluid outside the cells, contains the remaining 35 percent. Both ICF and ECF contain electrolytes—chemicals that dissolve in water to become positive and negative ions. Positive ions are known as cations, while negative ions are called anions. ECF is comprised of a salt and chloride solution—NaCl—that breaks down to become sodium cations (Na^+) and chloride anions (Cl^-). ICF is composed of potassium cations (K^+). Both Na^+ and K^+ are essential for neurons to conduct impulses throughout the body.

In order to understand the electrical changes, look at Figure 1.8. When a neuron is not carrying an impulse, it's considered in a state of **polarization** (A). This means that Na^+ is more abundant outside the cell, and K^+ is more abundant inside the cell. In this state, the neuron has a positive charge on the outside of the cell membrane and a negative charge inside. When an axon receives a nerve impulse, it releases the neurotransmitter acetylcholine (ACh). ACh diffuses across the synapse and bonds to the ACh receptors, located on the **sarcolemma**. The ACh makes the sarcolemma permeable to the Na^+ ions. These Na^+ ions then rush into the cell membrane and the K^+ ions rush out of the cell. The neuron is then in a state of **depolarization** (B) and the charges acting on the membrane are reversed. Therefore, the inside of the cell has a positive charge whereas the outside has a negative charge. In addition, the cholinesterase inactivates the ACh. This is when the impulse is generated and transmitted through the body to prompt a reaction.

In order for the neuron to receive another impulse, it must be **repolarized** (C) and returned to the state it was in before it received the stimulus. Re-

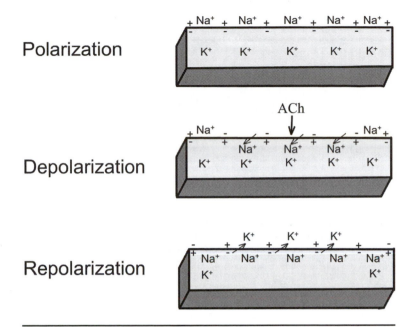

Polarization

Depolarization

Repolarization

Figure 1.8. Electrical charges.
Muscle contractions are caused by electrical charges and the ion chemistry at the sarcolemma. Polarization is when the muscle is relaxed, depolarization is when the muscle is responding to the influence of acetylcholine, and repolarization is the process of
the ion concentration returning to the polarization state. See also Table 1.1.

polarization happens when the body's sodium pumps return the Na^+ ions
outside the cell membrane, and the potassium pumps return the K^+ ions inside the membrane. The neuron is then once again in a state of polarization—the Na^+ ions are more populous outside and the K^+ ions are dominant
inside. The neuron is now ready to respond to a stimulus and generate an
impulse. Take a look at Table 1.1 for a recap of these processes. The polarization, depolarization, and repolarization cycle is similar to when a crowd
of cheering fans at a stadium sporting event begin a "wave." One section
will get up and throw their hands in the air, followed by the neighboring
section, and so on, and a wave motion reverberates around the stadium. Just
like the neuron's polarization behavior, the activity of each section influences the activity of the next section.

A neuron responds to a stimulus or action potential very quickly—in fact,
it's measured in milliseconds. Each neuron has the ability to respond to
hundreds of stimuli each second and generate an electrical impulse. Motor
neurons are especially efficient because only their nodes of Ranvier depo-
larize, which is known as salutatory conduction. In these neurons, impulses
travel extremely rapidly along the nodes of Ranvier. The neuron's myelin
sheath also increases the rate at which impulses are generated.

TABLE 1.1. Electrical Charges

Polarization (Resting Potential)

- A positive (+) charge is outside the sarcolemma and a negative (−) charge is inside.

- Na^+ ions are more abundant outside the cell. They diffuse inward until the sodium pump pushes them back outside the cell.

- K^+ ions are more abundant inside the cell. They diffuse outward until the potassium pump pushes them back inside the cell.

Depolarization (Action Potential)

- ACh is released by the neuron when a nerve impulse is received at the axon terminal.

- ACh makes the sarcolemma permeable to Na^+ ions.

- Na^+ ions rush into the cell, causing a reversal of charges on the sarcolemma—now the outside is (−) and the inside is (+).

- The ACh is then deactivated by cholinesterase.

Repolarization

- Cell becomes permeable to K^+ ions, which rush out of the cell.

- Charge restoration takes place—the outside is (+) and the inside is (−).

- The Na^+ ions return outside via the sodium pumps, and the K^+ ions return inside via the potassium pumps.

- Muscle fibers now are ready to respond to another nerve impulse received by the axon.

Once the impulse is generated, neurons are able to transmit the information within milliseconds, even over great distances. For example, when a barefooted man or woman over six feet tall steps on a sharp tack, they will feel **pain** just as quickly as someone who is only five feet tall. The body can transmit these sensory impulses from the sole of a foot to the brain in under a second.

The nerve fibers that actually transmit information to the spinal cord and brain are called primary afferent axons. The speed at which these fibers can transmit messages depends on their thickness—the thicker the fiber, the faster information can travel. These axons are classified into four different groups (in order of decreasing size—thickest to thinnest):

- *A-alpha*: carries information related to muscles

- *A-beta*: carries information related to touch

- *A-delta*: carries information related to pain and temperature

- *C*: carries information related to pain, temperature, and itch

Anesthetics

Most types of surgery require **anesthetics**. From major surgery, such as a heart bypass, to minor surgery, such as having a tooth removed, the operating doctor will administer some type of medication so patients will not feel pain. This can mean putting the patient "under," or causing them to go to sleep, or simply numbing the area that will be operated on.

Anesthetics are chemical solutions that block ions from passing through membranes. Because these ions cannot pass through the membrane walls, nerve impulses will not be conducted. Therefore, the impulses that would eventually transmit pain are diverted so they never reach the brain.

For example, if someone stubs her toe on a table, she initially feels the touch sensation when her toe collides with the table. The thick (and therefore fastest) A-beta nerve fiber carries this sensation from her toe to the brain. Because of the speed of this fiber, this sensation reaches the brain first. But shortly after, she feels pain. This is because the information related to pain is carried by the slower and thinner A-delta and C-nerve fiber.

The Spinal Cord

The central nervous system (CNS) is made up of two major organs: the brain and the **spinal cord**. The spinal cord connects the brain to the peripheral nervous system (PNS)—the nerves associated with the brain and spine. In order for information to travel between the brain and PNS, it must pass through the spinal cord.

Protected by a bony canal, the spinal cord extends down to the end of a column, which is made up of bones called vertebrae. This vertebral column grows at a faster rate than the nerve tissue of the spinal cord, so eventually the lower part of the canal grows longer than the cord. In an adult body, the spinal cord ends a short distance from the lowest rib (between the first and second vertebrae). An adult male's spinal cord will measure about 45 cm long, and an adult woman's will grow to about 43 cm long. The nerves that are below the spinal cord run in the vertebral canal—they are known as the *cauda equina,* which is Latin for "horse tail."

FUNCTION

The spinal cord has three main functions:

- *Direct reflexes*: Examples of a **reflex** are a kneejerk or pulling one's hand away from a hot surface. A **spinal reflex** only passes through the spinal cord and doesn't involve the brain.

- *Conduct sensory impulses*: Transmits sensory information from the afferent neurons to the brain through ascending nerve tracts.

- *Conduct motor impulses*: Transmits impulses from the brain through descending nerve tracts to the efferent neurons that communicate with the body's glands and muscles.

The **stretch reflex**—when a muscle is stretched and responds by contracting—is the most basic reflex arc in humans. This is basically a synapse between a sensory or afferent neuron and a motor or efferent neuron. Stretch reflexes can be induced by tapping many of the body's larger muscles, such as the triceps in the arm or the calf muscle in the leg. But most reflexes involve at least three neurons and numerous synapses. Reflex pathways or arcs will be examined more thoroughly later in this chapter.

STRUCTURE

Spinal Cord

The structure of the spinal cord is shown in Figure 2.1, a drawing of a cross-section or slice of the spinal cord. The center of the spinal cord contains H-shaped material—this is the gray matter, which is made up of the cell bodies of the motor neurons and interneurons. Surrounding this gray matter and filling out the remainder of the spinal cord is the white matter—consisting of the interneurons' myelinated axons and dendrites. These are known as nerve fibers, which are bundled into nerve tracts based on what function they perform. The descending tracts carry motor impulses away from the brain, and the ascending tracts carry impulses to the brain. Also note the **central canal**—this contains the **cerebrospinal fluid (CSF)**, which circulates in and around the brain and spinal cord.

Spinal Nerves

The spinal cord contains thirty-one pairs of spinal nerves (see Figure 2.2). Because they are in pairs, they are bilateral, which means the nerves and nerve tracts occupy both sides of the spinal cord. Whereas the spinal cord is considered part of the central nervous system (CNS), the spinal nerves and nerve tracts are part of the peripheral nervous system (PNS).

As noted in Figure 2.2, each pair of spinal nerves is numbered corresponding to the level of the cord and area of the vertebrae where it is attached. In addition, each pair is defined by a letter. There are eight **cervical** pairs,

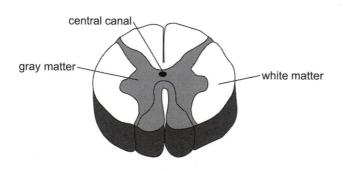

Figure 2.1. The spinal cord.
This cross-section of the spinal cord shows the gray matter, which contains the cell bodies of the motor neurons and interneurons, and the white matter, which contains the interneurons' myelinated axons and dendrites. The central canal contains the cerebrospinal fluid. The gray matter coming out the top of the figure is the dorsal root, and the gray matter coming out the bottom is the ventral root.

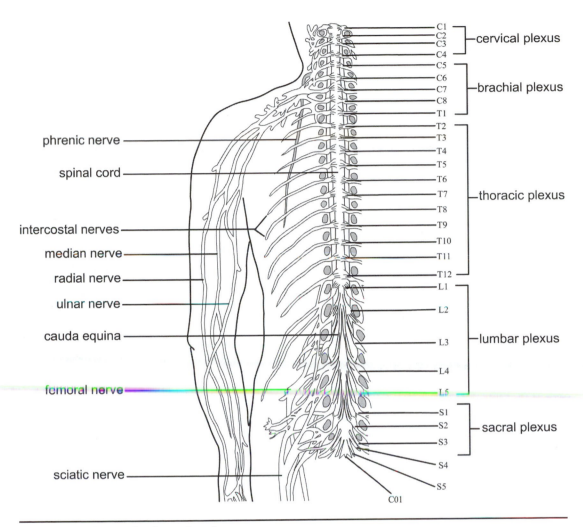

Figure 2.2. The spinal cord and nerves.

The CNS's spinal cord consists of spinal nerves (on the left of the figure) and the spinal nerve plexus network (located on the right of the figure). The purpose of the nerve plexus is to combine neuron networks from various parts of the spinal cord to form nerve systems specific to parts of the body. For example, the arms' radial and ulnar nerves are part of the brachial plexus.

twelve **thoracic** pairs, five **lumbar** pairs, five **sacral** pairs, and one tiny **coccygeal** pair (see Table 2.1). The first cervical vertebra is called the atlas, inspired by one of the heroes in Greek mythology. After losing an important battle, Atlas was turned to stone and forced to carry the Earth and heavens on his shoulders. Therefore, this vertebra is called atlas because it carries the weight of the head.

Receptors located on the skin send information to the spinal cord through the spinal nerve. In referring to Figure 2.3, notice the **dorsal** and **ventral**

TABLE 2.1. Spinal Nerves

Nerve Group	Name	Corresponding Location
C1–C8	Cervical pairs	Neck
T1–T12	Thoracic pairs	Ribs
L1–L5	Lumbar pairs	Large vertebrae in the small of the back
S1–S5	Sacral pairs	Base of the spine
CO1	Coccygeal pair	Pelvic floor

roots protruding from the spinal cord. These roots attach the nerves to the cord. Dorsal refers to the back or posterior of the body, and ventral refers to the front of the body. The dorsal horns are the gray matter in the dorsal area—this is where sensory information is received through the dorsal root. The ventral horns are the gray matter in the ventral region, and they contain the motor neurons. The axons of these motor neurons leave the spinal cord, travel along the ventral roots, and head directly to the muscles.

Each dorsal root contains an enlarged area of gray matter, called the **dorsal root ganglion**. Because ganglion refers to any collection of nerve cell bodies outside the CNS, the dorsal root ganglion contains the cell bodies of the peripheral sensory neurons, such as the receptors on the skin. All the nerve fibers from all the sensory receptors throughout the body converge in the dorsal root ganglia.

The ventral roots contain motor nerve fibers, which connect to voluntary muscles, involuntary muscles, and glands. Cell bodies from these neurons are housed in the gray matter of the spinal cord. Both the dorsal and ventral roots meet in the spinal nerve—therefore creating a network of both sensory and motor nerves.

In Figure 2.2 are groupings of nerves on the right side of the body called **nerve plexuses**. A nerve plexus is a combination of neurons from various sections of the spinal cord that serve specific areas of the body (see Tables 2.2 and 2.3).

Differences among Spinal Cord Segments

All the pictures in Figure 2.4 represent a different segment of the spinal cord. Each picture is slightly different. The darker areas in each picture represent the gray matter, which is where the cell bodies of the nerve fibers are located, and the white matter is represented by the lighter areas that surround the gray matter. This is where the spinal cord's axons are located.

In the cervical segment picture, there is a large amount of white matter. The cervical vertebrae fall right below the skull, at the top of the spinal cord.

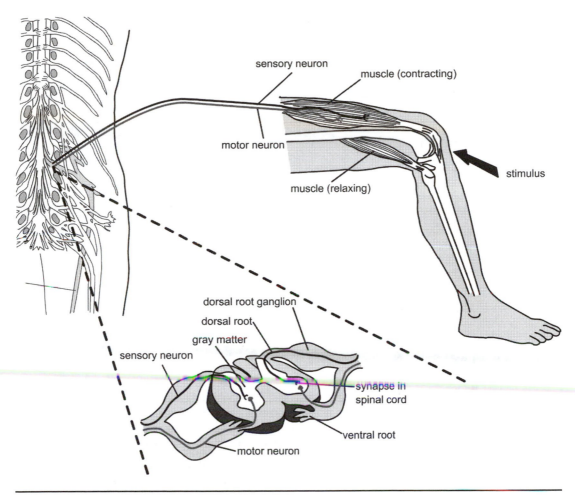

Figure 2.3. The spinal roots of the reflex arc.
A stimulus is applied to the knee, causing the sensory neuron to transmit the impulse to the spinal cord. The impulse is processed, and the instructions for a reaction are carried by the motor neuron, telling the relaxing muscle to contract. This reaction then causes the contracted muscle to relax.

Because of their location, there are many axons traveling up to the brain from all levels of the spinal cord. In addition, there are many axons traveling from the brain to the various segments on the spinal cord. In contrast, the sacral segment (which is the lowest, except for the coccygeal pair) has much less white matter. This is because there are fewer axons traveling to and from the brain through this spinal cord segment. To summarize, the amount of white matter in relation to gray matter decreases as one moves down the spinal cord.

In addition to the differences in the amount of white matter, there are also differences in the size of the ventral horn depending on the level of

TABLE 2.2. Peripheral Nerves

Spinal Nerves	Peripheral Nerve Group	Area Nerves Affect
Phrenic	C3–C5	Diaphragm
Radial	C5–C8, T1	Fingers: thumb, index, and middle; skin and muscles of posterior arm, forearm, and hand
Median	C5–C8, T1	Skin and muscles of anterior arm, forearm, and hand
Ulnar	C8, T1	Fingers: little and ring; skin and muscles of medial arm, forearm, and hand
Intercostal	T2–T12	Abdominal muscles; intercostal muscles; skin of body trunk
Femoral	L2–L4	Skin and muscles of anterior thigh, medial leg, and foot
Sciatic	L4–S3	Skin and muscles of posterior thigh, medial leg, and foot

TABLE 2.3. Nerve Plexuses

Plexus	Function
Cervical plexus	Contributes motor impulses to neck muscles; receives sensory impulses from the neck and the back of the head. The phrenic nerve comes from this plexus.
Brachial plexus	Supplies nerve branches to the shoulder, arm, forearm, wrist, and hand. The radial nerve comes from this plexus.
Thoracic plexus	Supplies nerves to upper and middle areas of the back.
Lumbar plexus and sacral plexus	Sends nerves down to the lower parts of the body. The sciatic nerve begins in this plexus—it begins at the dorsal part of the pelvis, passes through the gluteus maximus muscle, and extends down the posterior thigh. Near the knee, it divides to supply the leg and the foot.

the spinal cord (see Table 2.4). Motor neurons are abundant and large in the segments controlling limbs, which are the lower cervical (C5–C8), lumbar, and sacral sections. In addition, the thoracic level features an extra cell column called the intermediate or **intermediolateral cell column**. This cell column is the location of all presynaptic sympathetic nerve cell bodies.

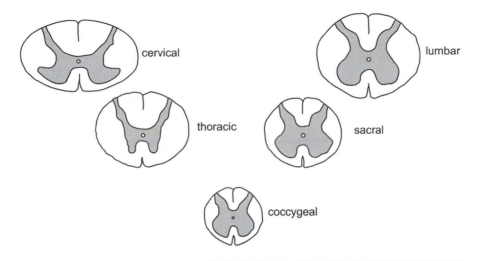

Figure 2.4. Spinal cord segments.
There are different amounts of gray and white matter at each level of the spinal cord. The cervical segment contains a large amount of white matter, whereas the coccygeal contains significantly less white matter and more gray matter in comparison. Because the cervical segment is high on the spinal cord and closer to the brain, it contains a large number of axons going to the brain from all spinal cord levels, in addition to going from the brain to the lower spinal cord segments. The coccygeal segment is the lowest level of the spinal cord and furthest away from the brain; therefore it has few axons traveling to and from the brain. The gray matter indicates a difference in motor neurons contained in the segments of the spinal cord. For instance, the cervical and lumbar segments both have enlarged ventral horns, which are the segments' lower halves. This is because they contain motor nerves that control the upper limbs, such as the arms (the cervical segment), and the lower limbs, such as the legs (the lumbar segment).

REFLEXES

As mentioned earlier in this chapter, a reflex refers to when an incoming signal is processed by the CNS and then reflected to the motor nerve fibers, which then generate movement. But the action is generated from the spinal cord and the brain is not directly involved.

The path that a nerve impulse follows after a signal is processed is known as the reflex arc. Five essential elements are involved in the reflex arc:

- **Receptors** detect the incoming stimulus and generate an impulse.
- Impulses are transmitted from receptors to the CNS through sensory neurons.
- The CNS houses the synapses where the impulse travels through.
- Impulses are transmitted from the CNS to the effector by motor neurons.
- The effector performs its distinctive action.

TABLE 2.4. Levels of the Spine

Level of Spine	Characteristics
Cervical	Wide, flat spinal cord; lots of white matter; ventral horn enlarged
Thoracic	Pointed tips between dorsal and ventral horns
Lumbar	Round cord; ventral horn enlarged
Sacral	Small, round cord; almost no white matter

An appropriate example of a reflex arc is the patellar reflex, also known as the kneejerk reflex. This is often an initial clinical test to determine if there is any neurological damage in the CNS. Doctors will hit the patellar tendon (right below the knee) with a rubber mallet to ensure a patient's nervous system is working correctly. If the knee quickly rises in response to the stimulus, then the CNS is functioning properly. Any problems with this response might indicate trouble in the thigh muscle or spinal cord. When the leg is raised, the muscle stretches and contracts, this is known as the stretch reflex.

In order to understand the patellar reflex, it's helpful to look at the different elements, which collectively happen in under one second (see Figure 2.3). When the stimulus, or rubber mallet, hits the patellar tendon, the **stretch receptors** detect that the tendon is stretching. These receptors produce impulses that are carried along sensory neurons to the spinal cord. In the spinal cord, the sensory neurons synapse with the motor neurons, which transmit an impulse to the motor neurons in the femoral nerve. These neurons in the femoral nerve then transport impulses back to the quadriceps femoris (known as the effector), which contracts and then causes the lower leg to extend.

A closer look at a reflex, such as the patellar or stretch reflex, reveals the important role of the **muscle spindle**, which is a small group of muscle fibers wrapped in connective tissue that separates it from the rest of the muscle. Connected to an afferent neuron, the muscle spindle actually detects the stretch in the muscle.

Another important spinal cord reflex is the flexor, or withdrawal reflex. Once again, the flex reflex is automatic and the brain is not directly involved in any decision making. This is when the stimulus is potentially harmful, such as touching a hot cooking pan or a sharp needle. The response is to pull one's hand or finger away. Similar to the patellar reflex, the sensory neurons transmit information to the spinal cord, and then the motor neurons tell the specific muscle to contract.

The Brain

Along with the spinal cord, the **brain** is the other major organ that makes up the central nervous system (CNS). Weighing approximately 3 pounds, or a little over 1 kilogram, in an average adult, the brain consists of over 100 billion neurons and trillions of **glia,** or support cells. The brain is covered by fluid, membranes, and bones. Housed in the **skull**, the brain is enclosed by a total of fourteen bones, eight of which make up the **cranium** (the remaining six bones enclose various other parts of the brain). The skull is important because it protects the brain, and it will be explained more thoroughly later on in this chapter.

The brain is made up of many major parts that function as an integrated unit, as shown in Figure 3.1. The **brain stem** is the first major part. This consists of the medulla, pons, and midbrain, which is located just above the medulla. The remaining major parts are the cerebellum, the hypothalamus, the thalamus, and the cerebrum. Although each part is explored individually in the following sections, it's important to remember that each part is interconnected and works with each other.

In addition to these major parts, the brain consists of four cavities, or ventricles, as noted in Figure 3.2. There are two lateral ventricles, as well as a third and fourth ventricle. Within each ventricle is a **choroid plexus**, a capillary network that forms cerebrospinal fluid (CSF) using blood plasma. This tissue fluid circulates in and around the brain and spinal cord, and will be covered in more detail later in the section about blood circulation in the brain.

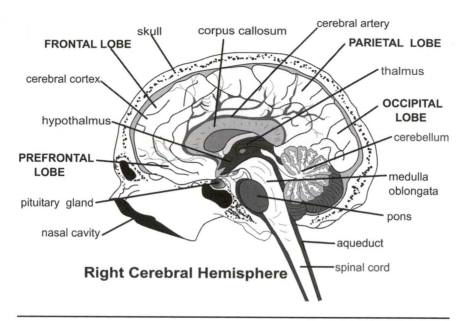

Figure 3.1. The right hemisphere of the brain, including the lobes.
This hemisphere of the brain is associated with space-related perceptions, facial recognition, visual images, and music.

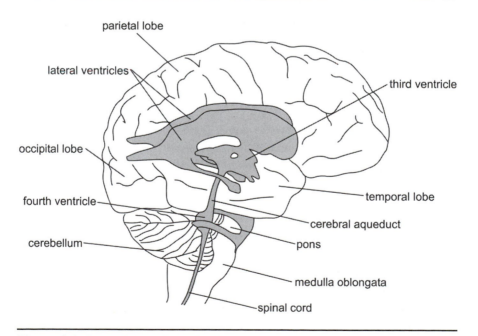

Figure 3.2. Ventricles of the brain.
The brain's four ventricles are cavities that contain the choroids plexus, which is important in forming the CNS's tissue fluid (called the cerebrospinal fluid).

MEDULLA

The **medulla** is anterior to the cerebellum, and extends between the spinal cord and the pons. Because the medulla contains myelinated nerve fibers and gray matter (or collections of cell bodies), this part of the brain is active in regulating the respiratory and cardiac activities of the body. This means regulating the body's heart rate, regulating the blood pressure by controlling the diameters (or width) of the blood vessels, and maintaining respiratory centers that control breathing. In addition, the medulla contains reflex functions that control coughing, sneezing, swallowing, and vomiting.

PONS

Also anterior to the cerebellum, the **pons** bulge out from the top of the medulla. The pons are also composed mainly of myelinated nerve fibers, which connect the two halves of the cerebellum with the brain stem in addition to the cerebrum above and the spinal cord below. This area of the brain is a vital link between the cerebellum and the rest of the nervous system. Nerve fibers in the pons carry messages to and from areas below it and above it. The pons contain two respiratory centers that work with the medulla to establish a normal breathing rhythm.

MIDBRAIN

The **midbrain** extends from the pons and encloses the **cerebral aqueduct**, which is a tunnel that joins the third and fourth ventricles. The upper part of the midbrain is composed of four rounded masses of gray matter. Visual and auditory reflexes—or eye and ear reflexes—are housed in these masses. For example, when a bee comes buzzing toward someone's nose, he moves his head away. This is a visual reflex because it involves the coordinated efforts of his eyeballs. When someone whispers near a person's ear, her first instinct will probably be to move closer to that whisper in order to hear better, which is an example of an auditory reflex. Righting reflexes—which ensure the head is upright and balanced—are also contained in the midbrain.

CEREBELLUM

As you can see in Figure 3.2 the **cerebellum** is separated from the medulla by the fourth ventricle and cerebral aqueduct and is located below the **occipital lobes** of the cerebrum. As stated earlier, the cerebellum is connected to the brain stem, cerebrum, and spinal cord by the pons. This part of the brain controls movement, which includes coordination, regulation of muscle tone, and maintaining posture and equilibrium in the body. Equilibrium is actually controlled by receptors located in the inner ear (see Chapter 5). It's important to note that these are all involuntary functions. Because the cerebellum works below conscious thought, the conscious brain is able to function without being overwhelmed. For example, if someone drops a pen while writing, the cerebellum coordinates the impulses that tell her arm, hand, and fingers to pick up that pen. This all happens unconsciously so the brain can focus on tasks that need conscious effort, such as reading or writing.

HYPOTHALAMUS

Located at the base of the brain (above the pituitary gland and below the thalamus), the **hypothalamus** acts as the junction between the nervous and the **endocrine system**. The hypothalamus is extremely small—it only comprises approximately 1/300 of the total brain weight. One of its most important jobs is to integrate all the various functions of the autonomic nervous system, which maintains the behavior of organs such as the heart, blood vessels, and intestines. But the hypothalamus has many other important and diverse jobs:

- *The body's thermostat.* The hypothalamus senses changes in body temperature and then transmits information so the body can adjust accordingly. For example, the hypothalamus will detect an increase in body temperature, which means the body is too hot. A signal will be sent to the skin's capillaries, telling them to expand to allow the blood to cool faster. Responses such as shivering or sweating are also prompted by the hypothalamus.

- *Oxytocin and antidiuretic hormone (ADH) production.* ADH helps to maintain the body's blood volume by enabling the kidneys to reabsorb water back to the blood. Oxytocin is important for women when they are in labor—it causes contractions that bring about delivery.

- *Stimulates the anterior pituitary gland.* The hypothalamus produces **growth hormone–releasing hormone (GHRH)**, which stimulates the anterior pituitary gland to secrete **growth hormone (GH)**. GH helps to encourage body growth throughout childhood—especially in bone and muscle development. But GH is also important in adults because it processes fats for energy production, increases the rate of cell division and protein synthesis, and speeds up the transportation of amino acids to cells, which enables protein production in the body.

- *Produces hunger sensations.* One of the hypothalamus's important jobs is to sense changes in blood nutrient levels. When these levels are low, this means that a person needs to eat to replenish nutritional resources. The hypothalamus activates this hunger sensation, so people eat and blood nutrient levels are raised. This creates the sensation of fullness, so that a person stops eating.

What Does the Brain Do while Asleep?

The average human adult sleeps 8 hours a day, 56 hours a week, 224 hours a month, and 2,688 hours per year. By the time a person is 75 years old, he would have spent a total of 25 years just sleeping.

Scientists can study brain activity during sleep by attaching electrodes to a person's scalp and then hooking these electrodes up to a machine called an electroencephalograph. The record of the brain activity is known as an encephalogram or EEG. An EEG is recorded in wavy lines, also referred to as brain waves.

Basically, there are two forms of sleep—slow wave sleep (SWS) and rapid eye movement (REM). There are actually four stages of SWS, which are followed by REM. Babies will spend approximately 50 percent of their sleeping in SWS and 50 percent in REM, and adults spend about 20 percent in REM and 80 percent in SWS. REM decreases with age—elderly people spend less than 15 percent of their total sleep time in REM. Dreaming occurs during REM. This is also the stage when the eyes move back and forth rapidly.

During sleep, the brain cycles through the four stages of SWS and REM about four or five times. For example, as a person drifts off to sleep, she enters SWS stage one. After a few minutes, she enters SWS stage two, and then three, and then four, which is followed by a period of REM sleep. Then she is in SWS stage four, then three, then two, and then one, and back again until the cycle escalates back up into REM.

Scientists believe that sleeping is important for restorative and adaptive reasons. The restorative theory is that humans and animals need to recover from all the work they've done throughout the day. Studies have shown that people who are deprived of any REM sleep are anxious, tired, and irritable the next day, and are unable to function at their optimum level. Many scientists believe that sleep developed as an adaptive mechanism. When humans were hunters and gatherers, they looked for food and water during the day. At night, however, they slept to store up their energy and hid to avoid getting eaten by enemies they couldn't see because of the darkness. In addition, they stayed in one place rather than walking around in the dark to avoid accidents such as falling off a cliff.

- *Stimulates visceral reactions in emotional situations.* When one is feeling intense emotions, such as anger or embarrassment, the hypothalamus detects a change in the emotional state and prompts a response by the autonomic nervous system. People can't control these visceral responses, and scientific experts still do not fully understand the biological and neurological bases for emotional reactions.

- *Maintains circadian rhythms.* The hypothalamus regulates body rhythms, sleep cycles, and accompanying changes in mood and mental alertness. Our **circadian rhythm** and biological clock ensure that people are awake and alert during the day and tired at night. When people sleep, the hypo-thalamic biological clock is reset, and they are able to be awake for the day. If someone gets too little sleep, his clock might not be completely reset and he will feel tired the next day. For more information on why humans need sleep, see "What Does the Brain Do while Asleep?"

THALAMUS

As shown in Figure 3.2, the third ventricle passes through both the hypothalamus and **thalamus**. Located above the hypothalamus, the thalamus focuses on sensations. Almost all sensory impulses travel through the thalamus. With the exception of smell, the sensory impulses initially enter the brain through neurons in the thalamus. Eventually these impulses are transmitted to the cerebrum, where they are processed (which eventually leads to perception), but initially the thalamus categorizes and integrates these impulses. For example, when someone holds a glass of ice water, her body feels impulses related to coolness and the feel of the glass, including its texture and shape, which is perceived by sensory receptors in her muscles. She doesn't feel these sensations separately because the thalamus integrates them before sending them to the cerebrum, where they are interpreted and felt.

The thalamus can also block minor sensations, causing one not to be distracted while intensely focused on a particular task. For example, when someone is engrossed in a good book or television program, he might not notice someone coming into the room or speaking to him. The thalamus allows the cerebrum to focus on that book or television program by suppressing these distracting sensations.

CEREBRUM

The largest part of the brain, the **cerebrum** is divided into two halves called hemispheres, which are separated by a deep groove or longitudinal fissure (see Figure 3.3). The hemispheres are also connected by a bundle of nerves called the **corpus callosum**, found at the base of the longitudinal fissure. A band of 200 million neurons, the corpus callosum allows the right and left hemispheres to communicate with each other. The brain stem connects the cerebrum with the spinal cord. As stated earlier, it is also the general term for the area between the thalamus and the spinal cord, which includes the medulla and pons.

As shown in Figure 3.1, the outer layer of the cerebrum is called the **cerebral cortex**, which is a sheet of gray matter tissue, about 2 to 6 mm in thickness. The word "cortex" comes from the Latin word for "bark." The cerebral cortex is similar to the bark of a tree—it serves the same protective function. The gray matter is composed of the cell bodies of neurons, which carry

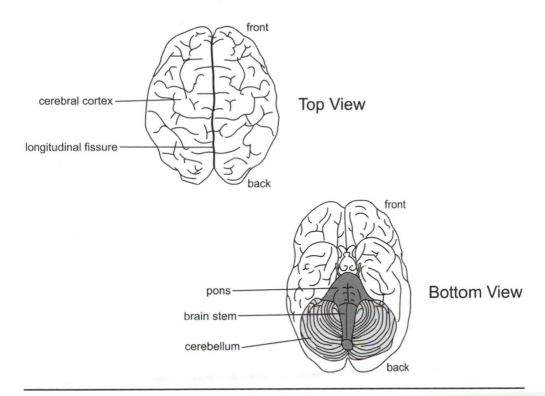

Figure 3.3. Top and bottom views of the brain.
The longitudinal fissure separates the brain's left and right hemispheres. The pons, which are visible in the bottom view, help to regulate breathing, and the cerebellum regulates movement. The brain stem contains the pons, along with the medulla and midbrain.

out the many functions of the cerebrum. White matter is also located inside of the gray matter. Consisting of myelinated axons and dendrites, the white matter connects the cerebrum's lobes to one another and to other parts of the brain (the brain's lobes will be discussed shortly).

The cerebral cortex is folded many, many times in the brain. In humans, the cerebral cortex looks like it has many bumps and grooves. These folds or bumps are known as convolutions or **gyri** (plural of gyrus), and the grooves between them are called fissures or **sulci** (plural of sulcus). Because of this extensive folding, millions and millions of neurons are located on the cerebral cortex. The degree or extent of folding corresponds with the brain's capabilities. In an animal such as a cat or a dog, the cerebral cortex does not have nearly the amount of folding as in a human; therefore, animals can not do many things that humans do—such as read, speak, or talk. See Table 3.1 and "Size Matters—Comparison of Brain Size of Animals and Humans."

Size Matters—Comparison of Brain Size of Animals and Humans

Just because a large animal has a large brain doesn't mean they are smarter. For example, an elephant's brain weighs 6,000 grams. In comparison, a human brain weighs 1,300–1,400 grams, but a human is obviously smarter. Large animals need bigger brains to control their large muscles and process the increased amount of sensory information due to their size. Take a look at Table 3.1 for the average brain size of some animals.

TABLE 3.1. Brain Size Comparisons

Animal	Brain Weight (g)
Elephant	6,000
Human	1,300–1,400
Monkey	97
Dog	72
Cat	30
Rabbit	10
Owl	2.2

LOBES

Certain areas of the brain are associated with specific functions. In addition, the cerebral cortex is divided into lobes, which are also associated with certain brain activities. Each hemisphere is separated into a frontal lobe, parietal lobe, temporal lobe, and occipital lobe (see Figure 3.4). Take a look at Table 3.2 for a summary of each lobe's functions.

The **frontal lobes** contain the motor areas of the brain—it's here that impulses are generated for voluntary motor activity and movement. The left motor area (in the left brain hemisphere) controls movement on the right side of the body, and the right motor area (on the right brain hemisphere) controls movement on the left side of the body. When someone has a stroke, also known as a cerebrovascular accident, either (or both) of their motor areas might be damaged, causing muscular paralysis. If the right motor area is damaged, then the left side of the body might be paralyzed. In addition to movement, the frontal lobe is involved in reasoning, planning, emotions,

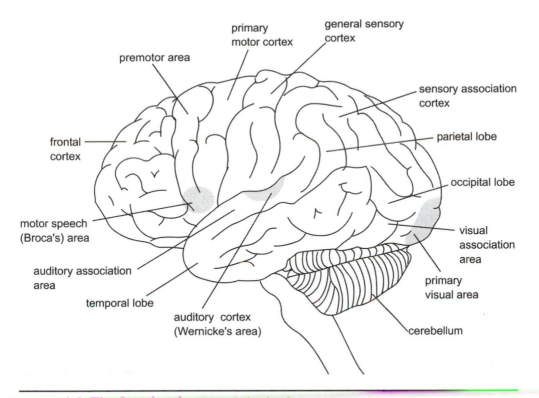

Figure 3.4. The functional areas of the brain.
These areas work to regulate speech, cognition, sensory association, motor abilities, and visual processes, among others.

TABLE 3.2. Lobes of the Brain

Frontal lobe	Motor areas: control movement
Parietal lobe	Sensory areas: perceive touch, pressure, temperature and pain
Temporal lobe	Auditory and olfactory areas: hearing and smelling; speech areas and hippocampus
Occipital lobe	Visual areas: vision, spatial distances

and problem solving. The frontal lobe in the human brain is relatively larger than in any other organism.

General sensory areas are located in the **parietal lobes**. This area focuses on receiving impulses relating to touch, pressure, temperature, and pain. Like the motor areas, the left area of the brain works with the right area of the body and vice versa. Skin receptors transmit impulses to the parietal lobes, where they are felt and interpreted. This area of the brain is also involved with stretch receptors in muscles and with taste—taste receptors, or taste buds, send their impulses to the taste areas of the parietal lobes. Taste areas overlap the parietal and temporal lobes.

Auditory (hearing) areas and olfactory (smelling) areas are located in the **temporal lobes**. Receptors for hearing are located in the inner ear, and olfactory receptors are located in the nasal cavity. In addition, temporal and parietal lobes in the left hemisphere contain speech areas that are involved with the actual thinking that precedes speech. Many people explain away an embarrassing slip of the tongue by exclaiming, "I spoke without thinking," but in reality, this is impossible.

The hippocampus, located in the temporal lobe and on the floor of the lateral ventricle, is involved with memory. Although little is known about how the brain actually stores and processes memories, it's believed that the hippocampus will collect information from many areas of the cerebral cortex, such as people's names or places where one has visited. If someone's hippocampus is damaged, they can only form memories that last a few seconds. This is evident in someone who suffers from Alzheimer's disease. Brain neurons are destroyed with this disease, followed by a loss of memory and the individual's personality (see Chapter 8). In addition to the memory functions, the hippocampus is also part of a group of structures known as the **limbic system**, which is important for controlling emotional responses, such as laughing and crying.

The last lobe area of the brain, the occipital lobe, is where the visual centers are located. Visual impulses received by the retinas in the eyes travel along the optic nerves to this area of the brain, where the brain processes and interprets what has been seen. Spatial relationships—such as judging distance and viewing in three dimensions—are processed in the occipital lobe.

Many areas in the cerebral cortex are not involved with movement or sensations. These are called association areas, and are believed to give people individuality and personality, including a sense of humor and the ability to learn and use reason and logic. See Table 3.3 for more details on the association areas of the cerebral cortex.

Two remaining important parts of the brain are the **basal ganglia** and the corpus callosum, which was mentioned earlier in this chapter. Grouped masses of gray matter within the white matter of the cerebral hemispheres, the basal ganglia regulates certain subconscious aspects of voluntary movement. Examples include movements such as hand gestures while talking or arm movements from front to back when walking. Parkinson's disease involves an impairment of the basal ganglia (see Chapter 8).

The corpus callosum is a band of nerve fibers that connects the left and right hemispheres of the brain and allows them to communicate and know

TABLE 3.3. The Cerebral Cortex

Cortex Area	Function
Prefrontal	Emotion, problem solving, complex thought
Motor association	Complex movement
Primary motor	Stimulates voluntary movement
Primary somatosensory	Receives tactile (touch) sensory information
Sensory association	Processes multisensory information
Visual association	Processes complex visual information
Visual	Receives simple visual information
Wernicke's area	Comprehends language
Auditory association	Processes complex auditory information
Auditory	Processes sound qualities, such as loudness or softness
Speech center, or Broca's area	Produces and articulates speech

This part of the cerebrum is in charge of many high-level functions, including language and learning, although language is managed in the left cerebral hemisphere. Figure 3.4 should be studied while reading the above breakdown of the cerebral cortex.

TABLE 3.4. Hemispheres of the Brain

Left Hemisphere	Right Hemisphere
Language	Spatial abilities and perceptions
Math	Facial recognition
Logic	Visual imagery
	Music

In general, each hemisphere of the brain is dominant for certain behaviors, although the information is shared through the corpus callosum.

each other's behavior. This creates a sort of "division of labor" in the brain because each hemisphere performs functions the other does not. For example, the left hemisphere contains the body's speech centers. Because the right hemisphere does not, it has to be able to communicate with the left hemisphere through the corpus callosum to know what these centers are "talking about." See Table 3.4 for a summary of each hemisphere's functions.

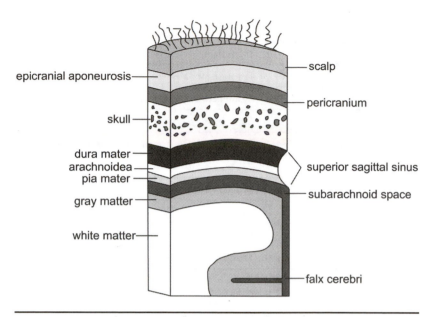

Figure 3.5. A cross-section of the skull and brain.
The scalp, pericranium, and skull all form the brain's coverings.

MENINGES

Because the brain is such an important organ, it needs a lot of protection. The initial layer of protection is the scalp, or skin layer, which contains the hair follicles. Right beneath the scalp is the skull, which is followed by three layers of connective tissues call **meninges** (see Figure 3.5). Meningitis—a nervous system disease involving the brain—is an infection of the meninges (see Chapter 8 for more information on nervous system diseases and disorders).

The outermost layer of the meninges is made up of thick fibrous tissue called the **dura mater**, which lines the skull. Because of its thickness, the dura mater keeps the brain from moving too much in the skull, which could cause blood vessels to stretch and tear. The next layer is the **arachnoid membrane**. Arachnids are spiders, so arachnoid is an appropriate name for this layer because it is made up of weblike strands of connective tissue. Finally, the innermost layer is called the **pia mater**, which is closest to the brain and spinal cord. An easy way to remember these layers is to: "The meninges PAD the brain"—P: pia mater, A: arachnoid membrane, and D: dura mater.

Between the arachnoid and the pia mater is the subarachnoid space, which contains the clear, colorless liquid called cerebrospinal fluid, or CSF.

CSF AND THE VENTRICULAR SYSTEM

Remember that the brain contains four **ventricles**, or cavities—the two lateral ventricles, and a third and fourth ventricle. Within each ventricle there is a choroid plexus, a capillary network where CSF is formed from cellular secretions and filtration of the blood, also known as blood plasma.

CSF flows from the lateral and third ventricles through the fourth ventricle. Then it continues to the central canal of the spinal cord, then to the cranial and spinal subarachnoid spaces.

CSF is reabsorbed into the blood as more is produced. It is through this continuous process that it flows in and around the CNS. When CSF is in the cranial subarachnoid spaces, it is reabsorbed through the **arachnoid villi** into large veins within the dura mater called **cranial venous sinuses**. After this reabsorption, the CSF becomes blood plasma again. The rate of reabsorption usually equals the rate of production. Total daily production of CSF is normally about 13.53–16.91 fluid ounces (400–500 milliliters) and the total volume of CSF remains around 4.28–5.07 fluid ounces (125–150 milliliters).

The CSF is vital to the brain and spinal cord, and has many functions:

- *Protection.* CSF cushions the brain, especially if the brain is impacted by a blow to the head. However, the CSF can only provide so much protection—sharp or heavy blows will injure the brain.

- *Buoyancy.* Pressure at the brain's base is reduced because it is immersed in CSF, reducing the net weight of the brain from about 52.91 ounces to 1.76 ounces (1,500 grams to 50 grams).

- *Waste Product Elimination.* The CSF takes harmful substances and toxins away from the brain.

- *Hormone Transportation.* The CSF transports hormones to all areas of the brain.

BLOOD SUPPLY

Blood brings necessities such as oxygen, carbohydrates, amino acids, fats, hormones, and vitamins to the brain. In addition, blood removes carbon dioxide, ammonia, and lactate from the brain. The brain occupies about 2 percent of the total body weight in humans, but it receives about 15 percent of the blood supply. The brain has priority over all other organs in the human body for blood because it needs blood to survive. Cells in the brain will die without oxygen, and it's the blood vessels within and on the surface of the brain that transport food and oxygen. Blood vessels bring their goods to the brain holes in the skull called **foramina**. (See "Blood-Brain Barrier.")

Blood-Brain Barrier

Scientists believe that a **blood-brain barrier,** or BBB, exists in the CNS to keep the brain free of too much blood and foreign substances that might cause injury. Because the BBB is semipermeable, certain substances can get in, but the barrier keeps others from crossing. The smallest blood vessels in the body are called capillaries, which are lined with endothelial cells. There are small spaces between each cell in the endothelial tissue that enable substances to move in and out of the capillary.

In the brain, however, these endothelial cells fit tightly together and substances cannot pass into or out of the capillary. This is known as the BBB. One of the only molecules that can break through the BBB is one from barbiturate drugs. The BBB has two important functions: (1) it protects the brain from harmful substances, and (2) it maintains a consistent environment for the brain.

There are numerous ways that the BBB can be broken down. One is by high blood pressure or hypertension, which opens the BBB. Exposure to microwaves and radiation can also break down the BBB, in addition to any severe trauma to the brain. However, the BBB develops with age—a baby has a weaker barrier system than a teenager.

Two pairs of arteries supply blood to the brain: the **internal carotid** and **vertebral arteries** (see Figure 3.6). The right and left vertebral arteries join at the brain's base to form the singular basilar artery. At the base of the brain, the basilar artery joins the internal carotid arteries in a ring, called the **circle of Willis**. This is a safety feature of the brain—if one of these arteries gets blocked, the circle of Willis still enables blood to get to the brain. A "brain attack" or stroke can occur if the blood supply to the brain is blocked, which can cause paralysis or aphasia (loss of speech). See "What Happens When Someone Has a Stroke?" for a thorough explanation of a stroke.

MEMORY AND LEARNING

Memory is the mental ability to remember ideas and experiences. As stated earlier, scientists believe the hippocampus is responsible for forming, sorting, and then storing memories. The hippocampus is also responsible for connecting new memories with related ones, thus giving them meaning. For example, one might remember the first baseball game she played by associating it with the smell of the grass on the field, the feel of the uniform she was wearing, and the sound of the crowd cheering for her. Each of these is an individual memory that relates to an overall recollection of that particular experience.

The initial stage of memory processing involves recognizing visual and au-

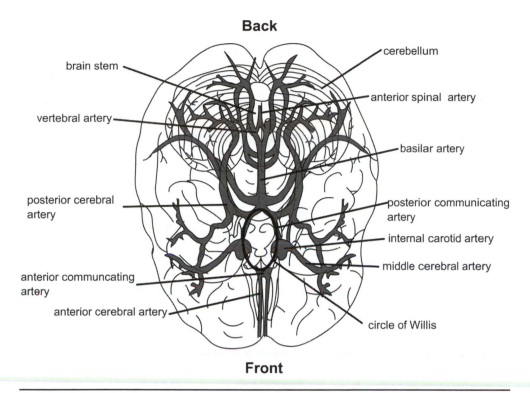

Back

brain stem

cerebellum

anterior spinal artery

vertebral artery

basilar artery

posterior cerebral
artery

posterior communicating
artery

internal carotid artery

middle cerebral artery

anterior communcating
artery

anterior cerebral artery

circle of Willis

Front

Figure 3.6. The arterial networks of the brain.
The posterior communicating artery, internal carotid artery, and middle cerebral artery form the circle of Willis, which provides the brain with blood if one of the internal carotid arteries or one of the vertebral arteries gets blocked.

ditory (sights and sounds) sensory signals that are stored for only a minute or fractions of a second. This **short-term memory** retains these bits of information only for a short time, and they are lost if not reinforced or repeated. When stored information is recalled at a later time, this is called **long-term memory**. Memories become more ingrained in the mind the more often they are repeated. Therefore, a short-term memory can become a long-term memory. In addition, the memory becomes stronger and more impressed in the brain the more often it is recalled—this is referred to as memorization.

For example, let's say someone is studying a history book on the Revolutionary War for an upcoming test. The more times he reads this book and processes the information, the easier it will be for him to recall what he reads when it comes time to take the test. Certain facts, such as dates of battles, will be imprinted on his memory; the deeper that imprint, the easier and quicker it will be for him to remember these facts. Studies have shown that someone who is wide awake and mentally alert memorizes far better than someone who is tired. Additional studies have shown that the brain is able to organize new

What Happens When Someone Has a Stroke?

Every year, 700,000 Americans suffer a **stroke**, and about 190,000 of these people die. A stroke, also called a **cerebrovascular accident (CVA)**, is when one of the brain's blood vessels is damaged, resulting in a lack of oxygen getting to that area of the brain. There are two types of blood vessel damage—**thrombosis** and **hemorrhage**.

A **thrombus** or blood clot often occurs as a result of **atherosclerosis**, which is when there are abnormal lipid deposits in the cerebral arteries. Because of these deposits, the surface of the arteries becomes rough rather than smooth. This obstructs the blood from flowing through the artery to the part of the brain it supplies. Thrombosis is when the blood clot occurs in the brain or neck, and a stenosis is when there is constriction of a blood vessel in the head or neck. Approximately 80 percent of strokes are caused by a thrombosis.

A hemorrhage results from an **aneurysm** of the cerebral artery. This causes blood to seep out into brain tissue, which puts excessive pressure on brain neurons, depriving them of oxygen and eventually destroying them. Symptoms of a hemorrhage begin quickly, and they include sudden weakness or numbness in the face, arms, and legs on one side of the body. If the stroke occurs on the left hemisphere of the brain, then the right side of the body can experience paralysis. Speech can also be damaged. When a stroke affects the vital centers in the medulla or pons, the patient can die.

For patients who suffer from the thrombus stroke, a clot-dissolving drug can help reestablish blood flow, although it has to be given within three hours of symptom onset. Recovery from both kinds of stroke depend on where it has affected the brain and the extent of the damage. It make take months of rehabilitation therapy for the patient to recover. One potential treatment is to tap into unused neurons. The cerebral cortex has many more neurons than the average individual uses, especially in those under 50 years old. One course of rehabilitation is helping the brain to find new pathways to work by making these little-used neurons work to their full potential.

information where similar information is stored. If, next month, that same person reads a book about George Washington becoming the first president of the United States, his memory will sort and file these new facts in the same area where it stored what he learned about the Revolutionary War. The memory will relate the new information on George Washington to what he learned about him during the Revolutionary War.

THE 10 PERCENT BRAIN THEORY

"The average human only uses 10 percent of his brain" is a scientific myth that many people have heard at one time or another. Even famous people such as Albert Einstein and Margaret Mead are attributed with stating this "10 percent myth." There are also advertisements on television or in the newspaper for medications or elixirs promising to unleash untapped resources in the human brain and transform people into smarter, more efficient people. But could this be true? If people only use 10 percent of their brains, then is 90 percent of the brain useless? Would anyone notice if it was removed?

The short answer is a resounding no—there is no scientific evidence to suggest people only use 10 percent of their brains. No one really knows how this 10 percent myth got started. Many historians trace the myth's origins to the early 1800s, when scientists were debating how the brain functions. Some thought that the brain acted as a whole (called equipotential), whereas others insisted that specific functions could be traced to certain areas of the brain. This second theory was related to the field of phrenology, which was developed by Franz Joseph Gall (1757–1828) and Johann Spurzheim (1776–1832). **Phrenology** was based on the idea that certain human behaviors and characteristics were dependent on the size, shape, and patterns of bumps on the skull. Critics of phrenology, namely Marie-Jean-Pierre Flourens (1794–1867), supported the equipotential point of view. They believed that although certain parts of the brain, such as the cerebral cortex, cerebellum, and hypothalamus, had specific duties, they all functioned as a whole.

In the late 1800s, a researcher named Karl Spencer Lashley (1890–1958) gained fame for his experiments analyzing rats' behavior when a portion of their brain was removed. Lashley believed that the rat only needed a certain amount of the cerebral cortex to function. In fact, his early tests showed that the rats performed well on visual recognition and maze tests even when fifty-eight percent of their cerebral cortex was removed. However, Lashley's

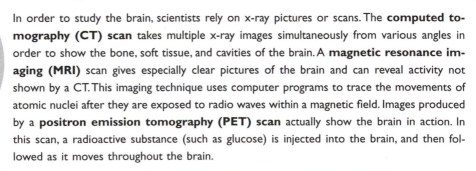

Taking Pictures of the Brain

In order to study the brain, scientists rely on x-ray pictures or scans. The **computed tomography (CT) scan** takes multiple x-ray images simultaneously from various angles in order to show the bone, soft tissue, and cavities of the brain. A **magnetic resonance imaging (MRI)** scan gives especially clear pictures of the brain and can reveal activity not shown by a CT. This imaging technique uses computer programs to trace the movements of atomic nuclei after they are exposed to radio waves within a magnetic field. Images produced by a **positron emission tomography (PET) scan** actually show the brain in action. In this scan, a radioactive substance (such as glucose) is injected into the brain, and then followed as it moves throughout the brain.

rats were never subjected to strenuous testing beyond simple tasks. Lashley only looked at the cerebral cortex of the brain and not the other areas, and the portions of the brain that he removed could have been used for tasks he never bothered to test. Nevertheless, it's possible that these results were exaggerated and interpreted to mean that only a small portion of the brain is ever used.

The weight of the average human brain is around 3 pounds (1,400 grams). If ninety percent of that were removed, we would be left with about 0.3 pounds (140 grams) of brain tissue, which is the size of a sheep's brain. Through medical research, we know that even damage to a small area of the brain, such as when someone suffers a stroke, can cause severe impairment. In addition, brain-imaging techniques such as positron emission tomography (PET) scans prove that the 10 percent myth is false. When these scans test for certain brain functions, such as visualization, it might appear that there are inactive areas of the brain. But in fact, these areas are active, just at a lower level compared to the active sites. (See "Taking Pictures of the Brain" for more on these scans.)

Peripheral and Autonomic Nervous System

The human body's nervous system is separated into two divisions: the central nervous system (CNS) and the peripheral nervous system (PNS). Chapters 2 and 3 explored the brain and spinal cord, the two primary components of the CNS. Through cranial and spinal nerves, the PNS transmits information to the CNS, where the brain processes it and responses are initiated.

The internal or **visceral organs** in the body, such as the heart and lungs, have nerve fibers and nerve endings that conduct messages to the brain and spinal cord. However, people are not aware that many of these messages reach their brain, although they do know that they happen or they wouldn't be functioning. In other words, these impulses never reach their consciousness. These impulses are processed or translated into reflex responses without ever reaching the conscious areas of the brain. For example, people do not notice when their blood vessels expand or their heart rates increase; they happen involuntarily.

These efferent or visceral neurons are grouped together in the autonomic nervous system (ANS), which falls under the direction of the PNS. This is where visceral neurons, which are neurons associated with the body's internal organs, relay information to the glands in addition to the smooth and cardiac muscles. Through nerve networks, the ANS facilitates communication between sensory impulses from the blood vessels, heart, and organs located in the chest, abdomen, and pelvis to various parts of the brain (especially the medulla, pons, and hypothalamus). Bypassing the consciousness, these impulses elicit mostly automatic reflex responses in the heart, vascular system, and bodily organs that control temperature, posture, food intake, and reactions to stressful feelings (such as anger or fear), among other processes.

CRANIAL NERVES

The ANS consists of twelve pairs of cranial nerves, which originate in the midbrain, the pons, and the medulla of the brain stem (see Figure 4.1) The pairs are numbered using Roman numerals—beginning in the front and ending in the back—that are based on the nerves' connection with the brain.

These cranial nerves are divided into sensory, motor, or mixed nerves depending on their function. Some of these nerves transport information from the sensory organs to the brain, and other cranial nerves work to control muscles. Some other cranial nerves control glands and internal organs, such

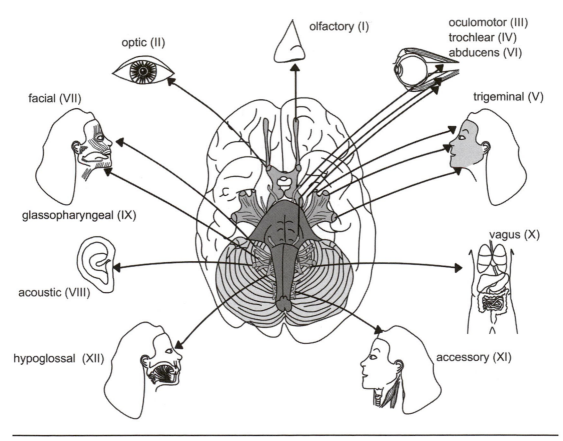

Figure 4.1. Nerve distribution.
The distribution of the brain's cranial nerve network.

as the ear and lung. Mixed nerves, which contain at least one sensory (or afferent) and one motor (or efferent) nerve, originate in more than one nucleus. In some cases, a single nucleus can produce more than one nerve. One example is the sense of taste, which comes from one nucleus even though its function is spread across two nerves.

Based on the sensory or motor functions of the cranial nerves, each pair is further defined by one of the following four categories:

1. *Special sensory impulses*: Senses relating to smelling, tasting, seeing, and hearing.

2. *General sensory impulses*: Senses relating to pain, touch, temperature, deep muscle sense, pressure, and vibrations.

3. *Somatic motor impulses*: These relate to voluntary control of the skeletal muscles.

4. *Visceral motor impulses*: These relate to involuntary reactions of glands and involuntary muscles such as the cardiac muscle.

Nerve pairs I, II, and VIII contain only sensory fibers, whereas III, IV, VI, XI, and XII contain mostly motor fibers. Pairs numbered V, VII, IX, and X contain mixed nerves (Figure 4.1 labels the twelve cranial nerves). Following is an examination of the cranial nerves based on their functions.

Optical sensations and eye muscles

Nerves III, IV, and VI are motor nerves that help the external eye muscles to operate. The **oculomotor nerve**, or Nerve III, controls blinking and pupil dilation. When one's pupils are dilated at the optometrist's office, the person receives eye drops, which are actually an acetylcholine blocker that inhibits the body's parasympathetic system, which will be explained later in this chapter. Once this happens, the pupils dilate, or expand, and the doctor can look inside the eye lens. However, the patient won't be able to read or focus on close objects because the parasympathetic system is being restrained.

The sensory aspect of this category involves Nerve II, the **optic nerve**, which transmits visual impulses from the eye to the brain. The other two cranial nerves in this category are motor nerves—the **trochlear** (Nerve IV) and the **abducens** (Nerve VI) control eyeball movement.

Face and mouth sensations

Nerve V is a mixed nerve that controls all sensations and movements from the face and mouth. For example, this nerve, also called the **trigeminal nerve**, controls chewing by carrying motor fibers from the face and mouth to the **mastication muscles**. Another example is getting a hard knot in the cheek when clenching the teeth. This happens when the trigeminal nerve stimulates the **masseter muscle**.

The sensory functions of this nerve include carrying impulses such as pain, touch, and temperature to the brain. For example, when someone is hit in the face, this nerve senses the onset of pain and changes in touch and transmits these impulses to the brain.

Another nerve in this category, Nerve VII, is simply known as the **facial nerve**. All of the muscles involved with facial expression are controlled by this nerve. However, the facial nerve is mixed because it also carries sensory impulses for taste from the tongue to the brain. Taste fibers, which originate from the taste buds, are predominantly located on the anterior two thirds of the tongue. However, the touch and pain sensations related to the tongue come from Nerve V.

Also classified in this category is the **olfactory nerve**, or Nerve I. This sensory nerve gathers sensations relating to smell from the nasal mucosa (the membrane located in the nose's interior that produces mucus) to the brain.

Hearing and balance

The **vestibulocochlear nerve**, or Nerve VIII, carries auditory or acoustic information relating to sound from the ear to the brain. In addition to hearing, this sensory nerve also controls balance.

Throat and salivary glands

Nerve IX, or the **glossopharyngeal nerve**, is a mixed nerve that contains sensory fibers for the throat and for taste from the posterior one third of the tongue. The tongue's movement is controlled by the **hypoglossal nerve**, or Nerve XII. Also located in this category are the motor nerves that control swallowing in the throat. In addition, this nerve also contains sensory fibers that control secretions from the salivary gland in the throat. The **spinal accessory nerve**, or Nerve XI, also controls the throat muscles, in addition to the two major neck muscles.

Thoracic nerve

Thoracic refers to the rib area of the spinal cord. Nerve X is the longest cranial nerve. Also known as the **vagus nerve**, it is a mixed nerve that controls most of the thoracic and abdominal organs, such as the glands, digestion (including the production of digestive juices), and the body's heart rate. In addition, the vagus nerve contains motor fibers that control the voicebox.

Table 4.1 contains a summary of each of the functions of the cranial nerves,

TABLE 4.1. Cranial Nerves

Pair Number	Name	Function	How to Test This Nerve
I	Olfactory nerve	Transmits smell impulses to the brain from the nasal mucosa (membrane that lines the nose and produces mucus).	Smell an odorous substance, such as a flower.
II	Optic nerve	Transmits visual impulses from the eye to the brain.	Follow your finger from side to side in front of your eyes; look down at the tip of your nose.
III	Oculomotor nerve	Controls eye muscle contractions, such as blinking, and pupil dilation.	
IV	Trochlear nerve	Controls eyeball movement.	
V	Trigeminal nerve	Main sensory nerve of the face and head; carries sensory impulses such as pain, touch, and temperature to the brain; also contains muscles for chewing, called muscles of mastication.	Touch your cheek; clench your teeth.
VI	Abducens nerve	Also controls eyeball movement.	Look to the side without moving your head.
VII	Facial nerve	Controls muscles for facial expressions; also contains sensory fibers for taste located on the anterior two thirds of the tongue, in addition to secretory fibers to the smaller salivary glands.	Smile; raise your eyebrows; taste sweet and salty substances.
VIII	Vestibulocochlear nerve	Also called the auditory or acoustic nerve; contains sensory fibers for hearing as well as balance.	Listen to an alarm on a clock; stand on one foot and maintain balance.
IX	Glossopharyngeal nerve	Contains general sensory fibers for taste from the posterior one third of the tongue; also contains sensory fibers for the throat (or pharynx) and secretory fibers that connect with the largest salivary gland. In addition, contains motor nerve fibers that control swallowing in the throat.	Place a depressor to the back of your tongue to test the gag reflex.

TABLE 4.1. (continued)

Pair Number	Name	Function	How to Test This Nerve
X	Vagus nerve	The longest cranial nerve, it works with most of the thoracic and abdominal cavities, such as the glands, digestion, and heart rate; also contains motor fibers for the voicebox or larynx, pharynx, and glands that produce various digestive secretions.	Open your mouth wide and say "AH."
XI	Accessory nerve	Also called the spinal accessory nerve, contains motor fibers with two branches; one branch controls the two neck muscles, and the other branch controls the throat muscles.	Raise your shoulders; turn your head from side to side.
XII	Hypoglossal nerve	Controls the tongue muscles.	Stick out your tongue.

including some sample tests a doctor might use to ensure these nerves are functioning properly. Failure of any of these tests might indicate nerve damage.

AUTONOMIC NERVOUS SYSTEM

An important part of the PNS, the ANS is made up of the motor elements of the cranial and spinal nerves. The ANS consists of visceral motor neurons that connect with smooth muscle, cardiac muscle, and glands. These areas are known as visceral effectors; they receive the impulses from the neurons through the nerve pathways and produce involuntary responses. For example, upon receiving impulses, the heart beats, muscles contract or relax, and glands secrete. Figure 4.2 depicts the ANS.

There are two divisions in the ANS: sympathetic and parasympathetic. In most cases, one will function in opposition to the other. The afferent nerves working for both divisions transmit impulses from sensory organs, smooth muscles, and the circulatory system, in addition to all the body's organs to the vital centers of the brain. Then efferent impulses are conveyed from these centers to all parts of the body by way of the parasympathetic and sympathetic nerves, which will be discussed shortly.

Basically, the sympathetic division operates in stressful situations and the parasympathetic controls the body in nonstressful situations. However, both divisions operate under the direction of the hypothalamus (Chapter

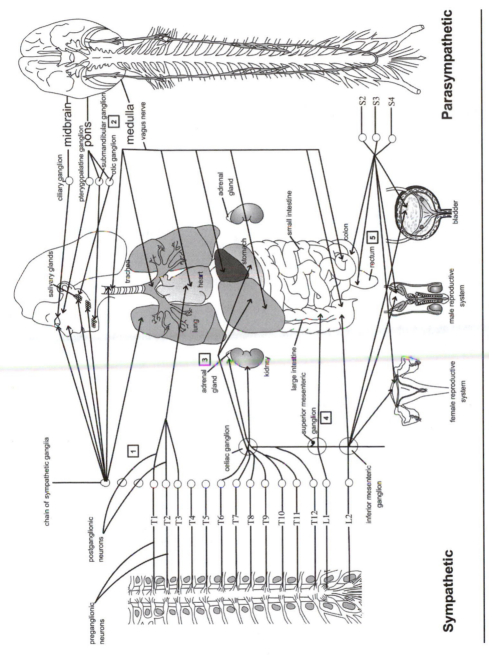

Sympathetic

Parasympathetic

Figure 4.2. The sympathetic and parasympathetic nervous systems.
The left side of the figure details the sympathetic system, along with the preganglionic and postganglionic neurons and the chain of sympathetic ganglia that connects the nerves to their effectors. On the right side is the parasympathetic system, which has various ganglia that connect the nerves to effectors.

3). As mentioned earlier, the hypothalamus is the part of the brain located right below the thalamus and above the pituitary gland. The hypothalamus has many functions, including regulating body temperature and ensuring that the visceral effectors (smooth muscle, cardiac muscle, and glands) respond appropriately to the impulses transmitted by the visceral motor neurons.

It's important to point out the role of autonomic pathways in the PNS. Impulses travel from the CNS along the autonomic nerve pathway to visceral effectors. Along this nerve pathway are located two motor neurons that synapse outside the CNS in a ganglion, which is a group of neuron cell bodies. The first motor neuron is called the **preganglionic neuron**, which connects the CNS to the ganglion, and the second neuron is called the **postganglionic neuron**, which connects the ganglion to the visceral effector. The ganglion is made of cell bodies of the postganglionic neuron. Fibers of the preganglionic neuron are short and myelinated, which means parts are covered with a myelin sheath that provides electrical insulation and increases the speed at which an impulse is transmitted. In contrast, fibers of the postganglionic neuron are long and unmyelinated.

These ganglia are like relay stations in the body. An impulse or message is transferred at the synapse (located in the ganglion) from the preganglionic neuron to the postganglionic neuron and then to the muscle or gland. This is in contrast to the voluntary nervous system, where the motor nerve fiber functions by beginning at the spinal cord and extending to the skeletal muscle without a synapse. It's through these processes of the autonomic pathways that stimuli is transmitted through the body, eliciting unconscious, automatic reflex responses such as the stomach and intestinal digestion, respiration rate and depth, pupil dilation, and blood rate regulation through the expansion or contraction of blood vessels.

SYMPATHETIC DIVISION

The cell bodies of the sympathetic division's preganglionic neurons originate in the thoracic and some of the lumbar segments of the spinal cord, which are located at the small (or lower area) of the back. Because of this, the sympathetic division is often called the thoracolumbar division. As shown on the left side of Figure 4.2, the ganglia of this division are located in two chains right outside of the spinal column. The sympathetic ganglia connect with them on these two chains. The ganglia is home to the synapses between the preganglionic and postganglionic neurons. After receiving an impulse, a preganglionic neuron synapses with a postganglionic neuron to initiate a response in an effector. It's important to note, however, that one preganglionic neuron often synapses with many postganglionic neurons, and thus many effectors. This allows the simultaneous responses among many effectors—a vital function of the ANS.

When a person becomes angry, afraid, or under any kind of stress, the sympathetic division of the ANS becomes dominant in the body (this also includes during exercise). This was essential for the survival of humans' ancient or prehistoric ancestors. Their lifestyle included hunting live animals for food and protecting themselves and their families from enemies such as animals and other humans. This demanded intense physical activity and endurance. This division of the ANS is often referred to as the "fight or flight response" which was obviously an important inner navigating tool for avoiding danger and defending family and property.

Even thousands and thousands of years later, the human nervous system isn't much different than that of its prehistoric forefathers. The fight or flight response still helps the body to determine the appropriate behavior when it feels afraid or anxious. Table 4.2 shows how some of the body's organs respond when in a stressful situation. The heart rate increases, and breathing becomes heavier because the bronchial muscles are contracting and the bronchioles are dilating, allowing for increased air intake. In other words, the breathing rate increases. A person might feel more powerful because the liver is changing glycogen to glucose, which supplies the body with more energy. Blood vessels associated with the visceral organs and the skin constrict, thus forcing more blood to vital organs such as the heart, muscles, and brain. But not all organs move at a faster rate. Stomach digestion is not important in stressful situations; therefore, secretion of digestive juices decreases along with peristalsis, which is the waves of muscle contractions that move food through the stomach muscle. All of these responses enabled prehistoric ancestors to stay and fight or run away from potential danger. People often find themselves in stressful situations that are not life-threatening, such as when they are studying for an important history exam or interviewing for a new job. But the body is prepared to react appropriately when circumstances escalate from everyday stress to life-threatening danger. See "How the Adrenal Medulla Helps Us Fight" to read how hormones enable the body to react appropriately in these situations.

An example of an everyday stress relieved by the sympathetic division is when the body temperature rises when one is sitting by the pool or on the beach. When someone is in a warm environment, this external stimuli can drain the body's heat reserves. Within the sympathetic division, thermal receptors send messages to the brain through the sympathetic nerve systems. One result of these messages is the expansion of **cutaneous** blood vessels, which reside right below the skin's surface. The expansion or dilation of these blood vessels enables more blood to flow to the body's surface where heat has been lost. This dilation also may cause oozing of certain fluids from

TABLE 4.2. Autonomic Nervous System: Visceral Effectors

Effector	What Happens When the Sympathetic Division Is Activated?	What Happens When the Parasympathetic Division Is Activated?
Eye muscle (also known as an iris)	Pupils dilate	Pupils contract
Salivary glands	Saliva production decreases	Saliva production increases
Nasal and oral mucus (mucosa)	Mucus production decreases	Mucus production increases
Heart	Heart rate increases	Heart rate decreases
Lungs	Bronchial muscle relaxes	Bronchial muscle contracts
Stomach	Peristalsis reduces	Peristalsis increases, gastric (or digestive) juices are secreted
Small intestine	Digestion processes slow down	Digestion processes increase
Large intestine	Movement and contractions slow down	Secretions increase, along with movement and contractions
Kidney	Urine secretion decreases	Urine secretion increases
Bladder	Organ wall relaxes and sphincter closes	Organ wall contracts and sphincter relaxes or opens
Liver	Glycogen changes to glucose	None
Sweat glands	Production of sweat increases	None
Skin and viscera blood vessels	Constrict	None
Skeletal muscle blood vessels	Dilate	None
Adrenal glands	Secretion of epinephrine and norepinephrine increases	None

If an organ receives both sympathetic and parasympathetic impulses, the responses are opposite. Note that some effectors can only receive sympathetic impulses (liver, sweat glands, many blood vessels, and adrenal glands). In these cases, an opposite response happens when there is a decrease in the sympathetic impulse.

the capillaries, causing dependent limbs to swell, which is why people sometimes swell in the heat. This increase in blood flow ensures that all necessary organs are receiving the proper amount of blood. Otherwise, the blood might concentrate in the lower limbs and not get to the brain, which leads to fainting spells.

Another important way the sympathetic division responds to excessive heat is by forcing the body to sweat. The brain's hypothalamus senses a great

How the Adrenal Medulla Helps Us Fight

The body's endocrine system controls glands that produce or secrete chemical messengers, known as **hormones**, into the bloodstream. One of these glands that is stimulated by the ANS is the **adrenals**. The adrenals are located right above the kidneys, and each is made of two small glands that operate independently. The inner area is called the medulla, and the outer area is called the cortex. The two main hormones released by the adrenal medulla, epinephrine (also known as adrenaline) and norepinephrine, are related to reactions to stressful situations also known as the "fight or flight response." Once these hormones are released, some of the effects include an increase in the heart rate, rate at which the body cells metabolize; dilation of the lungs' bronchioles, so the breathing rate can increase; and the conversion of glycogen (which is stored in the liver) to glucose. Through the blood, this glucose is sent to the voluntary muscles, enabling them to handle an increased workload.

increase in temperature and conveys this information to the sweat glands via the sympathetic nerves, which causes one to sweat. Through the evaporation of sweat (sometimes aided by a cool breeze), the body cools down. Once again, this is all involuntary. However, one can cool the body by jumping in a pool or cool shower or even sitting in an air-conditioned room, which lowers the environmental temperature.

PARASYMPATHETIC DIVISION

Another name for the parasympathetic division is the **craniosacral division**. In this division, the cell bodies of preganglionic neurons are located in the brain stem and sacral segments of the spinal cord. The axons of these preganglionic neurons are contained in cranial nerve pairs III, VII, and IX in addition to some sacral nerves. These axons extend to the parasympathetic ganglia, which are located in or extremely close to the visceral effector. The postganglionic cell bodies are actually located in the effector, and their very short axons connect with the cells of the effector.

Unlike in the sympathetic division, one preganglionic neuron synapses with only a few postganglionic neurons and then to only one effector. This aspect of the parasympathetic division enables single-organ responses, or very localized responses.

As stated earlier, this division dominates the body during nonstressful and relaxed situations, allowing several organ systems to function at a normal level and rate. For example, digestion is efficient through increased secretions and peristalsis. Urination and defecation occurs, and the heart rate will be at the normal resting rate.

NEUROTRANSMITTERS IN THE ANS

As mentioned, there are two kinds of synapses in the ANS that bring about a reaction in a visceral effector: one between preganglionic and postganglionic neurons, and another between postganglionic neurons and the effectors. In order for nerve impulses to cross synapses, they need the help of neurotransmitters.

Acetylcholine is the neurotransmitter released by all preganglionic neurons in both the sympathetic and parasympathetic divisions. A chemical inactivator is located at the dendrite of each postsynaptic neuron to defuse the impulse generated by the neurotransmitter. This is because transmitters must be controlled and inactivated—if their secretion was not stopped at some point, then rapid changes in excitation and inhibition would not occur and all activity would be slowed down. Acetylcholine's inactivator is cholinesterase, which is located in postganglionic neurons. However, in the parasympathetic division, postganglionic neurons also release acetylcholine right before they connect with their visceral effectors. In addition, most postganglionic neurons in the sympathetic division also release the neurotransmitter norepinephrine when they synapse with effector cells. Norepinephrine's inactivator is **catechol-O-methyl transferase**, or COMT. "Breakdown of Smooth Muscle Contractions" provides a detailed look at how the neurotransmitters work with the smooth muscles to contract.

AGING AND THE NERVOUS SYSTEM

Some typical nervous system problems that arise as one ages include dry eyes, constipation, and heart problems, which can be caused by a decreased stimulation of vasoconstriction, or blood flow, that is regulated by the sympathetic division.

Many believe that as the nervous system ages, the brain loses neurons, which leads to a decrease in mental capability and memory. This is not true. The brain does indeed lose some neurons, but the loss is only a small percentage of the total neurons. Some forgetfulness, as well as a decreased capacity for "quick thinking" or rapid problem solving, is to be expected, which is why older people generally need to drive slower and be more focused when operating a car. However, most memory capabilities remain intact in a healthy elderly person, and with the proper medical attention and care, most people can minimize these undesirable aspects of aging. The

Breakdown of Smooth Muscle Contractions

Muscle fibers are cylindrical cells that join together by the thousands to make up a muscle. The axon of this motor neuron contains sacs of the neurotransmitter called acetylcholine. This neurotransmitter's receptor sites are located on the muscle membrane, called the sarcolemma, which also contains acetylcholine's inactivator—cholinesterase. The gap between the axon terminal and the sarcolemma is called the synaptic gap or cleft, or synapse.

Muscle contraction actually begins with two proteins contained in the muscle fiber—myosin and actin. Each muscle fiber contains hundreds of sarcomeres, which are tiny units that contract. Sarcomeres are arranged end to end in cylinders called myofibrils. The center of the sarcomeres contains the myosin, and the actin is at the end of the cylinder. The interaction between myosin and actin causes the muscles to contract, and two other proteins—troponin and tropomyosin—are inhibitory. These proteins keep the myosin and actin in place when the muscle fiber relaxes. Outside of the sarcomeres is the sarcoplasmic reticulum, which houses an important component in muscle contraction, the calcium ions (Ca^{+2}).

The muscle contraction process goes through the same phases of electrical changes as a nerve impulse—polarization, depolarization, and repolarization (see Chapter 1). This process is called the sliding filament theory, and results in a muscle contraction. But it begins when a nerve impulse reaches the axon terminal and prompts the release of acetylcholine, which stimulates electrical changes and the movement of ions within the muscle fiber.

The sarcolemma is polarized when the muscle fiber is relaxed. This means that the outside of the muscle membrane has a positive charge and the inside has a negative charge. As you might recall from Chapter 1, this means the sodium ions (Na^+) are more abundant outside the cell, whereas the negative ions and potassium (K^+) ions are more abundant inside.

Because the Na^+ ions have the tendency to spread into the cell, the sodium pump transfers them back to the outside. The K^+ ions have the opposite movement tendencies—they tend to spread to the outside of the cell, and the potassium pump then has to return them to the inside. Therefore, the pump keeps the muscle fiber relaxed and polarized until the nerve impulse stimulates an electrical change.

A nerve impulse travels to the muscle fiber and goes straight to the axon terminal, where it prompts the release of acetylcholine. After spreading across the synaptic cleft, the acetylcholine bonds to receptors on the sarcolemma, which causes the membrane to be vulnerable to an incoming flow of Na^+ ions. This results in a reversal of charges, known as depolarization. Specifically, the rush of the Na^+ ions makes the inside of the sarcolemma positive, and the outside then becomes negative. Depolarization generates an electrical impulse or an action potential, which spreads along the muscle fiber. This impulse causes the myosin filaments to contract, pulling the actin filaments toward the center of the sarcomere, making it shorter, and forcing the muscle fiber to contract.

main causes of dementia or mental impairment are illnesses and disorders of the nervous system such as depression, malnutrition, and heart disease. Unfortunately, many disorders, such as Alzheimer's disease, are currently incurable. Nervous system diseases and disorders will be covered more comprehensively in Chapter 8.

5

The Senses

The central nervous system's five senses—seeing, hearing, touching, tasting, and smelling—allow the body to maintain homeostasis, which is when the internal environment is stabilized despite what constant changes are occurring in the external environment. The senses also protect people by providing information about what is going on inside and outside the body. For example, smelling and tasting might tell someone that something she is about to eat could be dangerous. Our touch sensation tells someone that a stove is too hot and will burn his skin on contact.

The information collected by the senses is transmitted through pathways and stimulates electrical nerve impulses. There are four important components of these sensory pathways:

- *Receptors.* Changes, or stimuli, are detected by receptors. All receptors respond to stimuli by generating electrical nerve impulses (see Chapter 1). However, depending on location, each receptor only responds to certain sensory changes. For example, receptors in the retina detect light rays while nasal cavity receptors detect airborne chemicals. When the specific stimuli is detected, an impulse is generated.

- *Sensory neurons.* These neurons take the impulses produced by the receptors and transmit them to the central nervous system (CNS). Although the sensory neurons are located in the spinal and cranial nerves, each carries impulses from only one type of receptor. For example, separate networks of these neurons serve the eyes, nose, ears, skin, and mouth.

- *Sensory tracts.* Impulses are transmitted to a specific part in the brain through sensory tracts in white matter located in the brain or spinal cord. In Chapter 1, white matter was defined as nerve tissue composed of myelinated axons and dendrites.

- *Sensory area.* This is where the impulses, or sensations, are felt or perceived and interpreted. Located in the cerebral cortex, this area functions without a person's conscious awareness.

BREAKDOWN OF SENSATIONS

Sensations have several important characteristics that enable people to feel, see, hear, smell and taste. The first characteristic is **projection**. When a hand pets a furry cat, it seems like the sensation is located in the hand. However, receptors located in the hand collect information associated with the cat's fur and transmit it to the cerebral cortex or the brain, where it is interpreted as soft and fluffy. The brain projects what it feels to the hand. This aspect is evident in patients who have had a limb amputated. Often patients say that even though their hand has been removed, they feel as if that hand is still there. Even though the hand's receptors have been removed with the severed limb, the nerve endings associated with those receptors still continue to generate impulses. The brain continues to behave as it did when the hand was still present. When the impulses from these severed nerve endings travel through sensory pathways to the brain, these impulses are interpreted and projected. The brain projects the sensation or feeling of the hand as still present. This feeling is known as **phantom pain**, and generally diminishes as the severed nerve endings heal.

Another important sensory characteristic is *intensity,* which is how strongly sensations are felt. A weak stimulus, such as a soft hum or dim light, will affect only a small number of receptors. However, a strong stimulus, such as a loud bang or bright light, will affect many more receptors, causing an increased amount of impulses to travel to the brain's sensory area. Based on the number of impulses received, the brain will respond accordingly. The more impulses received, the more intense the brain's sensory projection.

The brain's interpretations also allow it to contrast previous and current stimulations and allow for inflated or diminished sensations. For example, when someone takes a hot shower, the brain will compare the water temperature to those previously experienced. If the water is hotter than experienced before, the brain will most likely cause one to jump away from the water. But if the water is cooler than usual, the brain will tell one to make it hotter.

A third characteristic of the senses is **adaptation**, or when the body adjusts to a continuing stimulus. Receptors are always ready to detect changes to the body's external environment, but if the stimulus continues, it becomes less of a change. Therefore, the receptors will generate fewer impulses to the brain, which adapts itself to the stimulus. For example, many people wear jewelry on hands and arms, such as rings or watches. The presence of a ring is a continuous stimulus from the moment it's put on in the morning until it's taken off before bed. However, because it is on all day, the cutaneous or skin senses adapt to the presence of the ring and the wearer becomes unaware that it's on his finger. Only when there is a change, such as when the ring comes off before bedtime, do the receptors detect a change.

After-image is the final characteristic of the senses. This is when a sensation remains in the conscious memory even after the stimulus has ceased. One example is a flashbulb from a camera, which often stays in the memory for a few minutes after a picture is taken. Because the flashbulb produces such a bright light, the receptors in the retina generate many impulses that are interpreted by the brain as an intense sensation. The sensation is so strong that it lasts a bit longer than the stimulus from which it was generated.

CUTANEOUS SENSES

As mentioned earlier, **cutaneous senses** are those related to the skin. The cutaneous senses tell what is happening in and to the immediate external environment, including what is happening to the skin. For instance, an annoying mosquito bite on the knee often produces an itching sensation. This is actually a mild form of the pain sensation. The brain interprets sensory impulses in the parietal lobes. The largest part of this sensory area is reserved for the parts of the skin with the most receptors, which are the hands and face.

Sensations related to touch, pressure, pain, and temperature (heat and cold) are produced by receptors located in the skin's inner layer, or dermis (Figure 5.1). Pain receptors, also called free nerve endings, react to any intense stimulus. This means intense extremes of temperature applied to the skin, whether cold or hot, will be felt as pain. Encapsulated nerve endings are the receptors for the other cutaneous senses. This means that these nerve endings are surrounded by a cellular structure, as noted in Figure 5.1.

TASTE AND SMELL SENSES

Taste-specific receptors are located in taste buds, which are found in the **papillae** area of the tongue (Figure 5.2 outlines the mouth and nasal portions of the head involved with tasting and smelling). In general, experts believe there are four types of taste receptors: sweet, sour, salty, and bitter. The papillae's taste receptors (or chemoreceptors) decipher chemicals from food that have dissolved in the mouth's solution, also known as saliva. A moist mouth full of saliva is necessary for taste distinction—if the mouth is dry, even the most flavorful food, such as a grilled steak, will have little taste. Most foods contain a variety of the four general flavors and stimulate

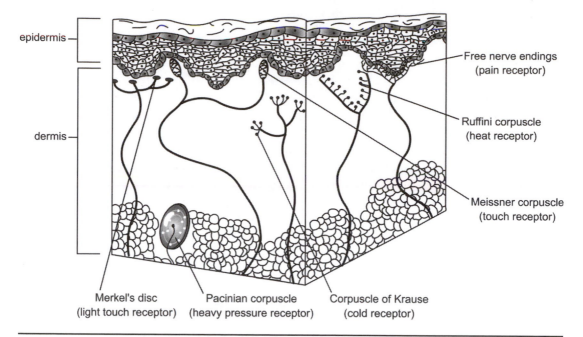

Figure 5.1. The cutaneous senses.
The cutaneous receptors of the skin include the Merkel's disc, the Pacinian corpuscle, the corpuscle of Krause, the Meissner corpuscle, the Ruffini corpuscle, and the free nerve endings.

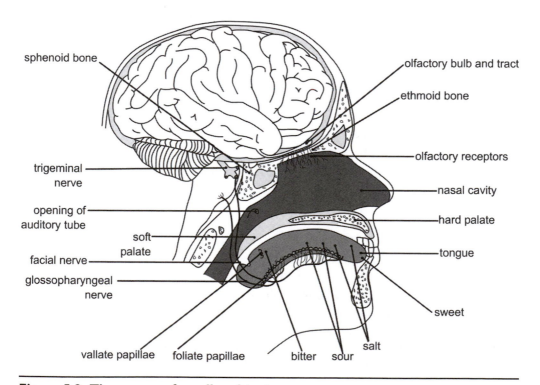

Figure 5.2. The senses of smell and taste.
This midsagittal section of the brain shows the smell and taste nerves and sensory components.

different combinations of receptors. For example, one bite of a peanut butter and jelly sandwich will stimulate both sweet (from the jelly) and salty (from the peanut butter) taste combinations. The smell of food also contributes to food perception.

Impulses relating to taste sensations are transmitted through the facial and glossopharyngeal nerves, which are the seventh and ninth cranial nerves, to the parietal-temporal cortex of the brain where the taste areas are located. Food scientists have found evidence of a genetic link in some taste preferences. For example, people who have an above average number of taste buds tend to find broccoli bitter and unpleasant, whereas those with an average number of taste buds might like this vegetable's taste.

Olfaction, or the sense of smell, functions through chemoreceptors located in the upper nasal cavities (Figure 5.2). These receptors detect vaporized chemicals and then generate impulses that travel through the first cranial or olfactory nerves on the ethmoid bone. From there these impulses move on to

Taste: a young girl enjoys an ice cream cone. © Skjold Photographs.

the olfactory bulbs and then on to the olfactory areas of the brain's temporal lobes. Scientists believe there are at least 1,000 different smells.

In comparison to other animals, humans have a poor sense of smell. For instance, dogs have a far greater sense of smell—they are believed to smell 200 times better than humans. The olfactory sense has a rapid adaptation rate, which is why pleasant smells tend to be acute and sharp at first but then quickly fade away. As mentioned above, the taste sense is greatly influenced by smell, which is why food can lose its taste when someone has a cold and the nasal cavities are clogged.

VISCERAL SENSATIONS

Visceral refers to anything that involves the body's internal organs, such as the glands and the smooth and cardiac muscle. **Visceral sensations** are the result of internal changes. Two important visceral sensations are hunger and thirst. The receptors that detect hunger and thirst are believed to be specialized cells located in the brain's hypothalamus. Hunger receptors function by detecting deficiencies in blood nutrient levels, and thirst receptors look for deficiencies in the body's water content, or the body fluid's water-salt proportion.

People are not conscious of the hypothalamus detecting hunger and thirst because the brain projects these sensations. Thirst is projected to the mouth and throat, which will feel dry because less saliva is produced. Hunger is projected to the stomach, which contracts and feels empty. Usually people satisfy both sensations by eating and drinking. However, if hunger is not addressed with food, eventually the brain adapts and the hunger gradually decreases in intensity. This is because even though blood nutrient levels will decrease and prompt the hunger sensation, these levels eventually stabilize as fat from certain body tissues is converted to energy. Once this stabilization occurs, there are few changes for the receptors to detect, and hunger diminishes. However, the brain does not adapt if the thirst sensation is ignored. The body has no ability to stabilize as the water content decreases. Without stabilization, changes and fluctuations continue, which receptors continue to detect. The thirst sensation increases, and dehydration may result.

VISION

Vision receptors are located in the eye, along with a refracting system that directs light rays to the vision receptors located in the retina. See Figure 5.3 for details on the various parts of the eye.

The eyeball is protected by eyelids and lashes. Eyelids are able to open and close over the eye because they are made of skeletal muscle. Eyelashes border the eyelids and keep dust and other debris from the eyelids. In addition, there is a thin membrane called the **conjunctiva** that lines the interior of each eyelid. Many eye infections are forms of **conjunctivitis**, in which the conjunctiva becomes infected and inflamed, making the eyes red and itchy.

Located on the upper outer corner of the eyeball are the **lacrimal glands**, which produce tears that cleanse the eyes and keep them moist. Tears are taken to the eye's anterior region through small ducts. Blinking spreads the tears and allows them to wash the eye. Composed mostly of water, tears also contain an enzyme called **lysozyme**, which prevents bacteria from producing on the eye's surface. In the outer portion of the middle of the eye are the superior and inferior **lacrimal canals**, which are ducts that transport tears to the **lacrimal sac**. Located in the lacrimal bone, the lacrimal sac leads to the **nasolacrimal duct**, which empties tears into the nasal cavity. This is what causes a runny nose when someone cries.

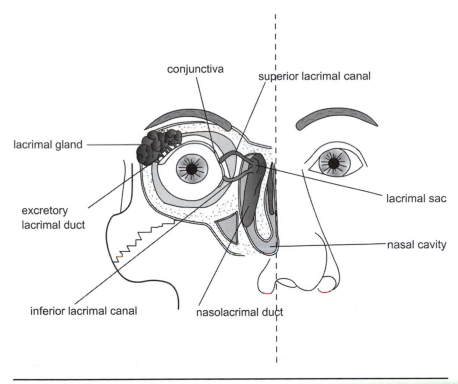

conjunctiva

superior lacrimal canal

lacrimal gland

lacrimal sac

excretory
lacrimal duct

nasal cavity

inferior lacrimal canal

nasolacrimal duct

Figure 5.3. The lacrimal apparatus.
The lacrimal apparatus is an important protection for the eye.

The **orbit** is a cavity in the skull that protects and surrounds the eyeball. There are six muscles (Figure 5.4) that extend from the socket to the surface of the eyeball. These six muscles are made up of four rectus muscles, which move the eyeball up and down and side to side. The remaining two oblique muscles allow the eye to rotate. These muscles function through the third, fourth, and sixth cranial nerve (oculomotor, trochlear, and abducens).

When examining the eyeball's anatomy (Figure 5.5), it's important to note the eyeball's three layers—the outer **sclera**, the middle **choroid layer**, and the inner **retina**. Composed of fibrous tissue known as the white of the eye, the sclera is the thickest layer. The **cornea** is located on the anterior of the sclera, and is unique from the rest of this layer because it is transparent and has no capillaries. This allows it to be the first part of the eye to bend (or refract) light rays.

The second layer of the eyeball, the choroid layer, is made up of blood vessels. In addition, this layer prevents glare by absorbing a certain amount of light within the eyeball. The outer portion of this layer contains the **iris** and the **ciliary body**, a circular muscle that is connected to the lens's edge by suspensory ligaments. Similar to the cornea, the lens is transparent and

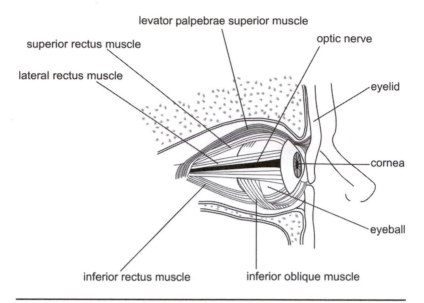

Figure 5.4. The muscles of the eye.
This view of the eye displays vital muscles, in addition to the optic nerve.

has no capillaries. The ciliary muscle allows the eye to focus light from objects near and far by changing the shape of the lens.

At the front of the lens the circular iris is located, which is known as the colored portion of the eye. The iris's opening is called the **pupil**. The pupil's diameter is controlled by two sets of muscle fibers. When the radial fibers contract, the pupil dilates or expands, which is a sympathetic response. The pupil constricts or reduces in size when the circular fibers contract, which is a parasympathetic response of the oculomotor nerves. This automatic or reflexive response is a protective mechanism because it prevents too much intense light from entering the retina. It also allows more precise near vision, which allows people to read books and other materials that are close to their eyes.

Another important part of the eye's anatomy is the retina, which is located on the interior of the choroid level (although it only covers two thirds of the eye). The retina house the visual receptors, called the **rods** and **cones**. Whereas rods only detect light, cones detect colors, which are actually made up of varying wavelengths of visible light. The **macula lutea** is abundant with cones and is located in the center of the retina behind the lens. The area known for the best color vision is the **fovea**, a small depression located in the macula lutea that contains only cones. Towards the edge of the retina is where the most rods reside. When light is dim, such as in a dark room or at night, we can best see through the periphery or sides of our visual fields, because this is where most of the rods are located.

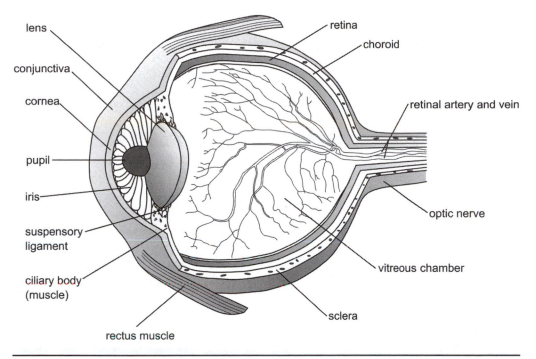

Figure 5.5. The anatomy of the eye.
The anatomical elements of the eye.

Inside the eyeball are two cavities. Located between the lens and the retina, the **larger posterior cavity** contains a semisolid substance called **vitreous humor**. The retina is kept in place by the vitreous humor. However, if the eyeball is injured and the vitreous humor is lost, the retina can become detached. The second cavity, the **anterior cavity**, is between the front of the lens and the cornea. This cavity contains the **aqueous humor**, the eyeball's tissue fluid formed by capillaries in the ciliary body. Aqueous humor flows through the eye's pupil and is absorbed by small veins called the **canal of Schlemm**, located where the iris and cornea join together. Because the lens and cornea have no capillaries, they need the nourishment from the aqueous humor.

HOW THE EYE PERCEIVES LIGHT

The vision process begins when the eye's receptors—the rods and cones—detect light and generate impulses. These impulses are then transported to the brain's cerebral cortex, where they are processed. An important part of the vision process is the **refraction** of light, which is when a light ray is bent or deflected as it passes from one object into another of greater or smaller density. In the eye, the refraction of light begins with the cornea, then the aqueous humor, the lens, and finally the vitreous humor. Adjust-

Why Do Some Eyes Need Glasses?

Vision problems are caused by errors of refraction. In addition, most vision problems, such as nearsightedness and farsightedness, are hereditary.

Normal vision is referred to as 20/20, which means that the eye can clearly see an object 20 feet away. If someone is nearsighted, or has **myopia**, the eyes can see near objects but not distant ones. For example, if someone has 20/80 vision, this means that the normal eye can see objects clearly at 80 feet away but the nearsighted eye can only see that object if it is brought within 20 feet. Focusing of images by the nearsighted eye is done in front of the retina because the eyeball is too long or the lens is too thick. This can be corrected with glasses with concave lenses that spread out the light before it hits the eye.

When an eye sees distant objects well, it is farsighted, or **hyperopic**. For instance, vision might be 20/10, which means that it can see at 20 feet what a normal eye can see at 10 feet. This eye will focus images behind the retina due to a short eyeball or too thin lens. To correct hyperopia, glasses with convex lenses are used to unite light rays before they hit the eye.

---

ments can only be made in the lens and depend on the **ciliary muscle**. When the eye is trying to focus on a distant object, the ciliary muscle will be re-

Elderly woman adjusts her glasses while reading. © Photodisc/Getty Images.

laxed, making the lens thin and stretched out. But when trying to focus on closer objects, the ciliary muscle contracts, causing the lens to recoil and bulge in the middle, which gives the lens greater refractive abilities. Problems with the lens and ciliary muscle can be addressed with corrective lenses or glasses (see "Why Do Some Eyes Need Glasses?").

The next step in the vision process happens when light hits the retina, causing chemical reactions to occur in the rods and cones. The rods contain a chemical called **rhodopsin**. During a chemical reaction, rhodopsin breaks down into scotopsin and retinal, a derivative form of vitamin A. An electrical impulse is generated as a result of this chemical reaction. The chemical reactions in the cones also involve reti-

Color and Night Blindness

When someone has **night blindness**, he or she cannot see clearly at night or when the light is dim. This can be caused by aging, or by a deficiency in vitamin A, which is used to synthesize rhodopsin in the rods. A lack of vitamin A will cause a decrease in rhodopsin, and the eye will not be able to process low levels of light.

Color blindness, a genetic disorder, occurs when one of the three sets of cones is dysfunctional. Many color-blind people can see at least some color—most have red-green color blindness and they cannot distinguish between these two colors. If either of these cones is missing, the eye can still see the color but will not be able to perceive the contrast; thus, the colors will look somewhat similar.

nal and generate an electrical impulse. However, cones are also absorbing various wavelengths of light during this time. There are three types of cones: red absorbing, blue absorbing, and green absorbing. Every ray of light is taken in by one of these types of cones. Dysfunctions of the cones and rods can lead to night and color blindness (see "Color and Night Blindness").

The impulses from the rods and cones are carried by **ganglion neurons** to the **optic disc**, where they converge to become the optic nerve and exit the eyeball. Because the optic disc contains no rods or cones, it is sometimes called the eye's "blind spot." However, the eye is constantly moving and rotating to compensate for this blind spot. The optic nerves from the left and right eye join together at the **optic chiasma**, located right in front of the pituitary gland.

Fibers from each eye's optic nerve cross to the other side, allowing each side to capture visual impulses from both eyes. In the brain, the visual areas are located in the occipital lobes of the cerebral cortex. These visual areas integrate the slightly different picture transmitted by each eye into a single picture, which is called **binocular vision**. In addition, the image on the retina is actually upside down, but these visual areas correct this so people see the image right side up.

HEARING

There are three main areas in the ear: the outer ear, the middle ear, and the inner ear (Figure 5.6). Receptors for hearing and **equilibrium**, or balance, are both found in the inner ear.

The **auricle** (or pinna) and **ear canal** make up the outer ear. Composed of skin-covered cartilage, the auricle is not important to humans, although it acts as a sound funnel for many animals, such as dogs and cats. Wearing

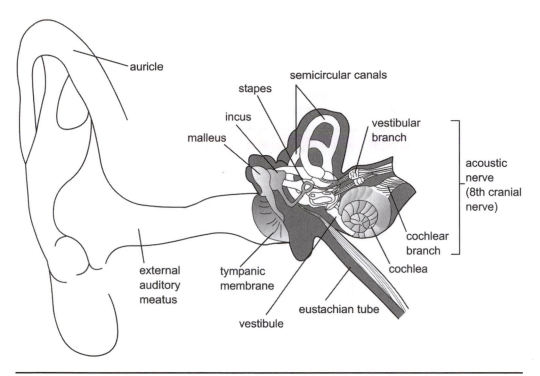

Figure 5.6. Inner ear structures.
The inner, middle, and outer ear structures.

glasses would be uncomfortable without an auricle, but it has no impact on hearing. The second part of the outer ear, the ear canal, acts as a tunnel into the temporal bone and middle ear. The **eardrum**, also called the tympanic membrane, stretches across the end of the ear canal and produces vibrations when hit with sound waves. These vibrations are transmitted to the three **auditory bones** called the malleus, incus, and stapes. This last bone, the stapes, transports vibration to the oval window through the inner ear.

Air enters and leaves the middle ear through the **eustachian tube**, also called the auditory tube, which extends from the middle ear to the nasopharynx. In order for the eardrum to function and vibrate, the air pressure in the middle ear must be equal to the pressure outside the ear. Ears often tend to "pop" when the air pressure is unequal, such as when in an airplane or when driving to a different elevation. The "popping" is caused when the eustachian tubes are trying to expand in order to equalize the pressure.

The inner ear is located within the temporal bone and encases a bone cavity known as the **bony labyrinth**. Lined with a membrane called the **membranous labyrinth**, this cavity contains fluid called **perilymph**, which is found between the temporal bone and the membrane. **Endolymph** is the fluid found within the inner ear structures—the **cochlea** (important in hearing) and the **utricle, saccule,** and **semicircular canals** (all important in equilibrium).

Figure 5.6 shows that the cochlea looks like a snail shell and is made up of two and a half turns that give it a coiled appearance. The cochlea is divided into three canals filled with fluid. The medial canal is known as the cochlear duct and encases the hearing receptors in the spiral organ, or **organ of Corti**. These receptors are known as hair cells, although they are not hair at all. These cells contain nerve endings from the cochlear branch of the eighth cranial nerve.

Hearing involves the reception of vibrations, the transmission of vibrations, and then the generation of nerve impulses. After sound waves enter the ear canal, they are transmitted to the ear structures according to the following sequence: eardrum, malleus, incus, stapes, the inner ear's oval window, the cochlea's perilymph and endolymph, and finally the organ of Corti's hair cells. Vibrations reach these hair cells, which bend and then generate impulses that travel to the brain through the eighth cranial nerve. Sounds are heard and processed in the auditory areas of the brain, which are located in the temporal lobes of the cerebral cortex. The hair cells are protected by the round window, which is found just below the oval window. The structure pushes out when the stapes pushes in through the oval window, thus relieving pressure and preventing damage to the hair cells. ("Deafness" contains an explanation of hearing loss.)

Two other inner ear structures, the utricle and saccule (see Figure 5.7), are located in the vestibule between the cochlea and semicircular canals. These structures are actually hair cells surrounded in a gelatinous membrane with **otoliths**, which are tiny crystals of calcium carbonate. When the head changes position, gravity pulls down on these otoliths and bends the hair cells, thus generating impulses. The vestibular portion of the eighth cranial nerve then transmits these impulses to the cerebellum, midbrain, and temporal lobes of the cerebrum. At the subconscious level, the cerebellum and midbrain interpret and process these impulses to maintain equilibrium. The cerebrum informs us of the head's position.

The last inner ear structure consists of three semicircular canals, which are also involved in stabilizing equilibrium. Each of these membranes is oriented in a separate plane and filled with fluid. At the bottom of each structure is the **ampulla**, an enlarged portion that contains hair cells sensitive to movement. When the body moves forward, the hair cells are initially bent backward and then straighten. Impulses are generated when these cells bend, and are also transmitted to the cerebellum, midbrain, and temporal lobes of the cerebrum via the vestibular branch of the eighth cranial nerve. The interpretation of these impulses are associated with stopping or starting, accelerating or decelerating, and changing directions. In

Deafness

There are three different types of **deafness** (the inability to hear properly): **conduction deafness**, **nerve deafness**, and **central deafness**. Conduction deafness is when one of the ear's structures cannot transmit vibrations properly. This can be caused by a punctured eardrum, an auditory bone arthritis, or a middle ear infection in which an excess amount of fluid fills the middle ear cavity.

Nerve deafness occurs when there is damage to the eighth cranial nerve or the hearing receptors located in the cochlea. Some antibiotics can damage this cranial nerve. In addition, some viral infections, such as mumps or rubella, can also cause nerve damage. Nerve deafness often occurs in the elderly when hair cells in the cochlea become damaged from years and years of exposure to noise; this hair cell damage is accelerated by chronic exposure to loud noise.

Central deafness is when the auditory areas of the brain's temporal lobes become damaged. This can be caused by a brain tumor or other nervous system disorder.

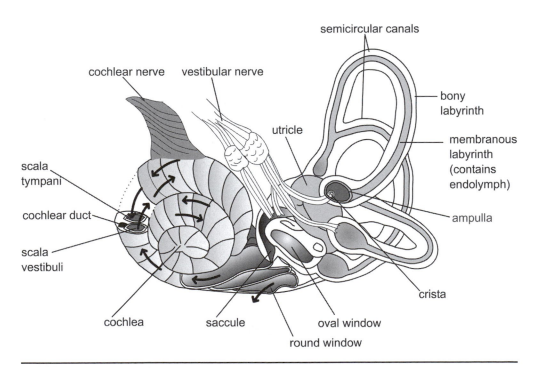

Figure 5.7. Detailed inner ear structures.
The inner ear structures are shown with details indicating the direction in which vibrations move during hearing.

general, the semicircular canals provide information while the body is in motion, and the utricle and saccule provide information while the body is at rest. The brain synthesizes all this information to create a unified sense of body position.

RECEPTORS IN THE BLOODSTREAM

Two of the heart's arteries, the **aorta** and **carotid**, have receptors to detect changes in the bloodstream. Blood pumped by the left ventricle is fed through the **aortic arch**, which works its way over the top of the heart. The **carotid arteries** are the left and right branches of the aortic arch that transport blood through the neck and then on to the brain (Figure 5.8).

Both of these blood vessel systems contain receptors. The **pressoreceptors** are located in the carotid sinuses, and aortic sinuses and detect changes in blood pressure; the **chemoreceptors** are located in the carotid and aortic bodies, and detect changes in the oxygen and carbon dioxide content of blood. Rather than stimulate sensations, these receptors generate impulses that regulate breathing and circulation. For example, if there is a sudden decrease in the blood's oxygen content (known as **hypoxia**), this change will be detected by the carotid and aortic chemoreceptors. These impulses are then transmitted through the ninth and tenth cranial nerves (glossopharyn-

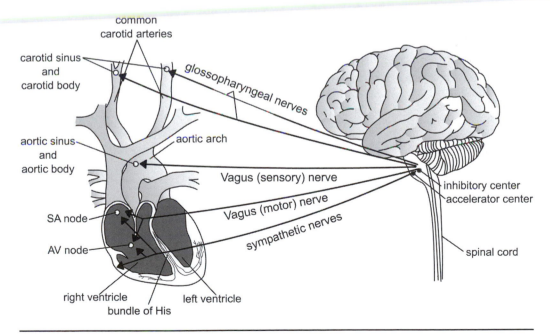

Figure 5.8. The regulation of the heart.
The brain and spinal cord work to regulate various elements of the heart.

geal and vagus) to the medulla. As a result of the sensory interpretation in the medulla, the respiratory rate and the heart rate will increase to collect and circulate more oxygen. This allows the body to maintain stable levels of oxygen and carbon dioxide in the bloodstream and maintain normal blood pressure.

EFFECTS OF AGING ON THE SENSES

As the body ages, each of the senses experiences some decrease in efficacy. For example, the eye lens can become opaque and form cataracts, causing diminished vision. In addition, the eyes might develop **presbyopia**, which is when the lens elasticity decreases and vision becomes more far-sighted.

The elderly are also at risk for glaucoma, an eye disorder that is caused by an increase in pressure on the eye due to an increase of aqueous humor in the canal of Schlemm. In a healthy eye, aqueous humor is produced and then reabsorbed into the canal of Schlemm, but in an eye with glaucoma, this reabsorption fails to take place, thus increasing the amount of the aqueous humor. The pressure from the aqueous humor is transferred to the lens and other parts of the eye, such as the retina. If left untreated, this pressure can impair vision or even lead to blindness.

Hearing can also diminish with age. After many, many years of noise, the hair cells in the organ of Corti become damaged. Usually deafness begins when high-pitched sounds are unable to be heard. The ears and brain may still be able to process low-pitched sounds, although eventually this can also decrease.

The sense of taste makes eating enjoyable, and people need to eat to survive. Unfortunately, this sense tends to diminish as people age. In addition, many elderly people are on medications that tend to diminish the taste abilities. This is one possible explanation for poor nutrition among elderly and sick populations.

6

History of the Discovery of the Nervous System

The history of the discovery of the nervous system is really the history of **neuroscience** (the prefix "neuro-" refers to the nervous system). Scientific historians and anthropologists trace the beginnings of the study of living things, which is defined as biology, to when *Homo sapiens* appeared over 50,000 years ago. During the Paleolithic Era, or Stone Age, these ancient humans were classified as hunters and gatherers (or hunter-gatherers) according to evidence uncovered from fossil records. During this time, humans began to distinguish themselves from their ape-influenced ancestors through the development of speech, abstract thought, and religion. They were also forced to confront the basic realities of existence, including disease, hunger, pain, birth, and death.

FROM HUNTERS TO GROWERS

About 10,000 years ago and following the Stone Age, humans focused on developing agriculture, including farming and animal husbandry techniques, rather than on hunting and gathering as a means of producing food. Because they were relying on nature's elements—wind, rain, sun, and water—in their food production, they grew preoccupied with planning and interpreting and understanding the world around them and their place in it. This preoccupation included their living and dying roles on Earth, which prompted an increased interest in health and medicine. Therefore, anatomy and anatomical study soon emerged as a primary goal of biological knowledge.

Ancient people had a great fear and respect for the spiritual world inhabited by the dead, which led them to conduct elaborate funeral proceedings when one died. Many ancient people felt that once death could be commemorated and even anticipated, attempts could be made to escape it through both rational and mystical means.

Some burial rituals included carving up the body and removing the organs, so the body could be preserved, or mummified. However, doctrines from various religious disciplines over time absolutely forbade the use of a knife on a dead body (or human dissection) for the purposes of anatomical investigation. Similar beliefs stated that the body belonged to God and not man, and therefore it was immoral to dissect a human body after death. Doctors in ancient Greece subscribed to this belief system and refused to conduct human dissections, thus limiting their exploration of the human anatomy. For hundreds and hundreds of years, scientists and researchers were forced to rely on animal dissections to gain an understanding of the human body.

GREEK SCIENTISTS

Intellectuals and philosophers of ancient Greece are generally credited with developing scientific study during the sixth century BCE. With a secular eye, these thinkers refused to accept spiritual or religious causes for the hows and whys behind the workings of humans and the natural world they inhabit. Although these questions were raised by earlier populations, the Greeks took them out of various religious boundaries and brought them into the realm of science and philosophy. However, it is important to remember that most of their anatomical studies were limited to animal models.

Democritus

One of the first important figures in the study of the human body's sensory mechanisms was a philosopher named Democritus (ca. 460–370 BCE). Not only a philosopher, Democritus is also considered an early atomist, because he believed the world and everything in it are composed of an incalculable number of atoms. The atoms exist and move around an infinite void of empty space, and in some manner make up everything in the universe— from people to animals to objects.

One of Democritus's most profound contributions was a primitive explanation of the five senses. He claimed that all things that were perceptible, meaning things that could be seen, tasted, smelled, felt, and heard, were distinct arrangements of atoms that differed only in size and shape. Human perceptions were the effect of the atom bodies on the sensory organs. For example, the interaction between the group of atoms making up an apple and its reaction with the space or air surrounding the apple creates the visual image.

Many consider Democritus one of the earliest anatomists because of his keen interest in animal dissections for the purpose of studying the body's organs. He is thought to have been one of the first scientists to classify animals according to two major categories—**invertebrates** and vertebrates. His philosophy and scientific knowledge led him to declare that the brain, rather than the heart, was an organ of thought, as was previously believed.

Hippocrates

One of Democritus's contemporaries, Hippocrates (ca. 460–370 BCE), agreed with this assertion and also claimed that the brain was the seat of consciousness and thinking. Regarded as the father of medicine, Hippocrates developed some medical foundations that doctors still rely on in modern society. Hippocrates believed that not only was the brain the conscious seat of the body, but it also controlled the senses and movement. Afflictions of the brain were the most challenging for a physician to treat—a belief that still resonates today. Physical ailments involving the brain, such as tumors and neurological disorders, continue to confound even the most brilliant doctors today (see "Egyptian Brains").

Aristotle

Another famous Greek philosopher and scientist, Aristotle (384–322 BCE) is well-known for his earth-centered view of the universe that dominated Western thought until Nicolaus Copernicus (1473–1543) introduced the heliocentric, or sun-centered, belief of the universe. But Aristotle had a great

Egyptian Brains

The oldest recorded use of the word "brain" appeared in an Egyptian document called the *Edwin Smith Surgical Papyrus*, which is believed to have been written in 1700 BCE. This papyrus is considered to be the first medical document ever written, and is thought to have been written by an Egyptian physician named Imhotep.

About 15 feet long and 13 inches wide, this papyrus describes forty-eight medical cases. In addition to mentioning the brain, the document discussed meninges (coverings of the brain), the spinal cord, and the cerebrospinal fluid, all for the first time in recorded history.

The surgical document was named after its discoverer, an American Egyptologist by the name of Edwin Smith (1822–1906). Smith bought the papyrus from a dealer when he was visiting Luxor in 1906. After he died, his daughter donated it to the New York Historical Society, and it was eventually published.

interest in medical and biological investigations, even if his conclusions and theories were erroneous. For instance, he contradicted Democritus and Hippocrates by declaring that the purpose of the brain was to secrete mucus to cool the body's blood supply by counteracting the great heat emitted by the heart. He also disagreed with Democritus by declaring that the heart was the body's intellectual center and soul and thus linked to one's emotional state. According to him, when the brain fails to cool the body, the heart causes the bloody supply to get hot, which in turn causes the person to become passionate and excitable; but when the brain is in control, the excessive cooling of the blood leads to sleep.

Galen

A few hundred years after Aristotle's death, another philosopher named Galen (129–199 BCE) contributed significantly to the anatomic understanding and the workings of the human body. His work remained the authority in the medical and biological worlds for hundreds of years, and his anatomical and physiological studies were not questioned until the fifteenth and sixteenth centuries. In fact, doctors in the nineteenth century who were attempting to reform various medical practices bitterly complained that some of Galen's flawed beliefs continued to have a stronghold on the scientific community. Due to religious dogma, Galen was forced to limit his dissections to animal models, rather than human ones, so it's not surprising that much of his research was later revealed to be defective.

Both Aristotle and Galen believed that the nervous system operated similarly to a set of hydraulic pipes, and they believed that the brain matter was actually less important than its ventricles, which were thought to be a fluid reservoir. However, Galen expanded upon this foundation, declaring that the human body was made up of three principal organs—the liver, heart, and brain—along with three types of vessels—the veins, arteries, and nerves. Ultimately, the body relied on air, or pneuma, which was processed by the liver into what he called "natural spirits," which were used to support the body's nutrition and growth functions. Galen believed that the brain received "vital spirits," which controlled movement, from the heart that mixed with the blood. The arteries distributed this vital spirit and blood mixture, which produced heat, in order to warm the body. Then the air was adapted by the brain to form "animal spirits," which governed sensation as well as muscular movement. Animal spirits were then distributed by the nerves.

Galen's anatomical progress is also demonstrated in his studies of the nervous system. He detailed the anatomy of the brain, spinal cord, and nerves and disproved Artistotle's view that the nerves originated in the heart by proving they were rooted in the brain and spinal cord. In a brilliant series of experiments, he explored spinal cord injuries. Galen found that when the spinal cord was injured between the first and third vertebrae, death would immediately result, and damage between the third and fourth vertebrae caused respiration to arrest (or breathing to stop). He also studied various forms of paralysis that resulted when the vertebrae and other areas of the spinal cord were damaged.

THE RENAISSANCE

Around the time of Galen's death, years of turmoil and warfare led to the collapse of the Roman Empire. This period, from about 500 to 1300, was known as the Middle Ages. Because the time of the Roman Empire (or the Roman Era) had been one of relative peace and prosperity, intellectual exploration and development were allowed to flourish, especially in the scientific realm. But with the political upheaval and religious conflicts that permeated the Middle Ages, many scientists were considered pagans and persecuted for pursuing their studies. It wasn't until the Age of the Renaissance (1300–1650) that interest in Greek ideals was reborn. Scientific study was again pursued, and thus the exploration of the human body and the nervous system continued.

One of the earliest Renaissance scientists was the Flemish physician Andreas Vesalius (1514–1564), who revolutionized the study of anatomy by pointing out various inaccuracies in Galen's work. Although both Galen's and even Hippocrates's contributions cannot be underemphasized, they were anatomically weak. Because human dissection was strictly forbidden in the Roman Empire, Galen was forced to settle for animal, rather than human, cadavers for his anatomical studies. He assumed that the essentials of animals and humans, such as their nervous systems, were more similar than different. Because Galen's beliefs dominated medicine and neuroscience for over 1,000 years, few dared to question him. But because Vesalius

Anatomy of the Brain (Paris 1546), Charles Estienne. © National Library of Medicine.

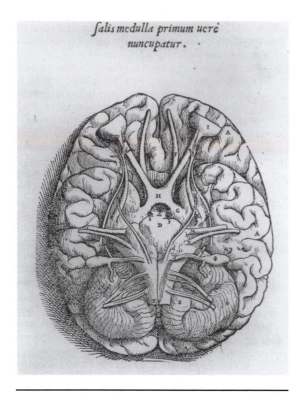

ſalis medulla ucrè nuncupatur.

Basal view of the brain and cranial nerves, *De Human Corporis Fabrica* (1543), Andreas Vesalius. © National Library of Medicine.

was able to examine human cadavers, he found erroneous conclusions in Galen's anatomy models, which Galen had applied to the human form based on careful dissections of animals such as monkeys, dogs, and pigs.

ARTISTS AND ANATOMY

Art during the Renaissance, such as that created by Leonardo da Vinci (1452–1519), is inextricably linked to anatomy. The Greek philosophers and scientists encouraged an accurate depiction of the human body and thus encouraged careful study and exploration. Unlike the Greeks, however, the Renaissance scientists and artists were often able to perform human dissections. Although quite rare, some even attended human dissections performed in front of an audience, which were called public anatomies. For example, Leonardo da Vinci is believed to have dissected as many as thirty human bodies throughout his studies.

AUTOMATIC AND PERISTALTIC REACTIONS

After the Greeks made great strides in exploring the mind and body relationship, scientists in the seventeenth century continued to discover how the nervous system operated. Many significant scientists during this time believed in the mechanist theory, which meant that the human body was a sum of its parts and that the various functions of the body could be investigated and described as if the body was a machine. One of the renowned scientists who applied the mechanist theory to the workings of the nervous system was the French physiologist René Descartes (1596–1650).

Descartes, who was also a famous mathematician and philosopher, is acknowledged as one of the initial leaders in the study of anatomy and physiology. His most significant contribution is distinguishing between voluntary and involuntary reactions. In 1633, he wrote a book called *De homme,* in which he outlined a proposed mechanism for the body's involuntary reactions, or the reflex arc. He explained that the body senses a force, such as when a hand touches the flame from a candle, which results in a pain sensation. The nerves transmit this sensation to the nervous system and then to the motor nerves exiting from the nervous system. Finally, the muscles that control action receive the pain message and cause the hand to pull away. Other automatic, or **peristaltic**, reactions that Descartes wrote about included intestinal contractions, such as during digestion.

Descartes also believed that the brain's pineal gland was the body's control center and controlled all of its actions. He thought the pineal gland produced a nervous fluid that commanded voluntary actions, such as muscle contractions.

Still a devout mechanist, Descartes thought the body operated like a well-oiled machine, with most functions operating automatically. Because he subscribed to Aristotle and Galen's beliefs that the nervous system behaved like a set of hydraulic pipes, he believed that a propulsion wave from the nerve's hollow interior composition enabled it to carry the stimulus to the brain. The nerves then expanded because they were full of this fluid, which caused the muscles to expand and ultimately contract. In *De homme,* he wrote,

> [O]ne can well compare the nerves of the machine that I am describing to the tubes of the mechanisms of these fountains, its muscles and tendons to diverse other engines and springs which serve to move these mechanisms, its animal spirits to the water which drives them, of which the heart is the source and the brain's cavities the water main.

INVENTIONS AND NEW TOOLS FOR DISCOVERY

In the mid-nineteenth century, scientists began studying the human body's nervous system by using a microscope. The invention of the microscope is widely attributed to Zacharias Janssen (1580–1638), a Dutch spectacle maker who introduced the concept of a compound microscope, which meant using a combination of more than one lens for magnification, during the late sixteenth century. But Janssen's model only provided ten times magnification. Shortly after Janssen's invention, a Dutchman by the name of Antony van Leeuwenhoek (1632–1723) and a British scientist by the name of Robert Hooke (1635–1703) independently refined these compound microscopes and produced greater magnification.

With this ability for greater magnification, Hooke and van Leeuwenhoek published their observations and illustrations on structures of microorganisms, including their cellular construction. Van Leeuwenhoek is also believed to be one of the first scientists to describe the details of a nerve fiber based on his microscopic observations. Hooke's book, entitled *Micrographia* (1665), is believed to feature one of the earliest mentions of the term "cell" in addition to illustrations of cellular structures. By looking at cork and plant sections under the microscope, Hooke compared the cork's "dead" cells (which appeared to be full of air) with the plant's living cells, which were oozing with fluids.

MOTION AND MUSCLES

About fifty years after Descartes wrote *De homme,* another mechanist researcher, an Italian physician by the name of Giovanni Borelli (1608–1679) published a landmark work on the physiology of motion, *De Motu Animalium.* In this book, Borelli focused on analyzing human motions such as running and walking and described muscles as similar to levers. Before Borelli's work was published, the Dutch anatomist Nicolaus Steno (1638–1686) stated that muscle action results from the movement of individual muscle fibers. Borelli supplemented Steno's work by suggesting that the nervous system feeds muscle fibers a chemical that causes a reaction within the fibers, thus leading those fibers to shorten and the muscle to contract.

Another physician who explored chemical reactions and the nervous system was Thomas Willis (1621–1675). In 1664, he published *Anatomy of the Brain, with a Description of the Nerves and Their Function,* which was the most comprehensive and accurate explanation of the nervous system at that time. His most important discovery was the identification of a hexagonal group of arteries located at the base of the brain. This artery system, which was named the circle of Willis, ensured that the brain received a sufficient blood supply at all times. Willis also identified the eleventh cranial nerve, also known as the spinal accessory nerve, which is responsible for stimulating movement of the major neck muscles. The discovery of this chemical reaction set the stage for understanding the role bioelectricity plays in the nervous system.

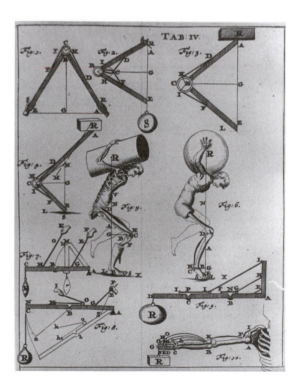

Illustration from Giovanni Borelli's book, *De Motu Animalium....* © National Library of Medicine.

Following the publication of Descartes and Willis's nervous system volumes, the next stage was set for the discovery of the role bioelectricity plays in the body.

THE DISCOVERY OF BIOELECTRICITY

Before Borelli analyzed the chemical nature of fluid coursing through the nervous system, great scientists had believed there was a "spiritual" element present in the body. In addition to believing the hydraulic model of the nervous system, Descartes and his contemporaries believed Galen's assertion that the fluid coursing through the body that motivated the reflex arc was known as vital spirits. For instance, Descartes thought that if the brain stopped receiving blood from the heart, nerve function would cease because the vital spirits could not travel to the ventricles for conversion into animal spirits, which then caused movement.

Many scientists and even philosophers speculated about the nature of these spirits, although few believed that humans could really generate electricity or electric currents in their bodies. In *Principia Mathematica* (1687), Sir Isaac Newton (1642–1727) said that all bodies contain a "certain most subtle spirit" and that all sensations are caused by excited vibrations of the spirit that travel along the nerves to the sensory organs, then to the brain, and from the brain to the muscles.

But it wasn't until the Italian anatomist and doctor Luigi Galvani (1737–1798) began conducting experiments using frogs that human electricity was given serious consideration. At the time it was known that certain animals, such as eels, generated electricity because they gave shocks when touched. Combining this with the belief that lightning was an electrical spark (as proposed by Benjamin Franklin [1706–1790] in 1752), Galvani suspended the legs of a frog with brass hooks from an electrical rail during a thunderstorm. He noticed that the muscles were continuously contracting—when the lightning appeared and when it didn't. Further testing led Galvani to place the frog on various different metals, indoors and outdoors, and he noticed that the frog's muscles continued to contract. Because the legs were contracting whether external electricity was applied or not, he concluded that the frog was generating electricity on his own. He said that certain organic tissues in the frog were capable of generating "animal electricity," a kind of energy he thought was similar but distinct from "natural" energy that was generated by machines or lightning. In 1791, Galvani published his conclusions in an essay entitled *Commentary on the Effect of Electricity on Muscular Motion*. In the essay, he drew a comparison between a muscle and an electrified jar. When an external charge was applied to the jar, opposite electrical charges on the inside and outside surfaces would result and cause an attraction. Similarly, when an external charge was applied to a muscle, opposite electrical charges would result in an attraction that led to a muscle contraction.

Galvani's experiments established the basis for the biological study of neurophysiology and neurology. In addition, Galvani finally toppled the long-standing hydraulic model of the nervous system, leading to an increased examination of nerve conduction. Nerves were not water pipes or channels, as Descartes and others had stated. Instead, they were electrical conductors. Galvani was the first scientist to propose that organic tissue carried information and messages throughout the nervous system using electricity.

NERVES

With the study of bioelectricity underway, scientists began focusing on the specific roles that nerves and nerve cells had in conducting information to the brain. In 1826, the German psychologist and physiologist Johannes Müller (1801–1858) proposed that different nerves were coded to carry out a specific function, which depended on where they originated in the brain. Known as the theory of **specific nerve energy**, Müller believed there were specific nerves for all of the five senses—sight, hearing, smell, touch, and taste.

Three other important anatomical discoveries followed on the heels of Müller's work. First, Robert Remak (1815–1865) suggested in 1836 that nerve cells and nerve fibers are joined, in addition to describing myelinated and unmyelinated fibers. Then, in 1837, Jan Purkyne (1787–1869) described cells in the brain's cerebral cortex. Both Remak and Purkyne wrote that the nerve fiber or axon arose from the nerve cell. Finally, in 1839, Thomas Schwann (1810–1882) proposed the cell theory, which stated that the nervous system was composed of hundreds and hundreds of neural cells. In this theory, he emphasized the role of the nucleus in cell formation and reproduction. A devoted disciple of Johannes Müller, he also discovered the sheath covering nerve fibers, which is named the Schwann cell in his honor (see "Nerve Impulse Is Measured").

Nerve Impulse Is Measured

Another one of Müller's students, Hermann von Helmholtz (1821–1894), is primarily well-known in the field of physics, although he was the first scientist to measure a nerve impulse. Through his invention of the myograph, von Helmholtz determined that the nerve impulse had a relatively slow speed of approximately 90 feet per second. This was a final blow to those believing in "vitalism" and the presence of a vital force that helped move vital spirits.

CONDUCTION

Although Galvani identified the connection between electricity and muscle contractions in the late 1700s, the technology was not yet available to measure this electrical current. After the invention of the **galvanometer** in the 1820s, a Swiss-German scientist by the name of Emil Heinrich Du Bois-Reymond (1818–1896) was able to detect and measure what he called an "action current" in the frog's leg in the mid-1850s. By connecting metallic electrodes from the nerve to the galvanometer, he stimulated the nerve in order to produce a muscular response. When this happened, Du Bois-Reymond noticed a small negative variation of the resting electrical potential in the electrodes, which he called "negative variation" and identified as the primary cause of a muscular contraction. This action current, which later became known as the **action potential**, was defined as an electrical impulse wave that traveled along the nerve fiber at a fixed speed.

By the 1830s, the Scottish surgeon and anatomist Charles Bell (1774–1842) had written extensively on the relationship between the special senses, their nerve networks, and where they could be traced from specific locations in the brain to their end organ (research initiated by Müller in 1826). Bell was one of the initial anatomists to clearly demonstrate that spinal nerves carry both sensory and motor functions. He also established that sensory fibers run along the posterior roots of the nerves and that the motor fibers run through the anterior (these views later became known as Bell's Law). In addition, Bell carefully studied nerves related to facial muscles. He proved that cranial nerve V's sensor components were related to the feeling in the face and that the motor components controlled chewing. According to Bell, cranial nerve VII controlled muscles related to facial expressions. Facial paralysis, due to a lesion on the facial nerve that results in distortion, was eventually named Bell's palsy in his honor.

Nerve damage was also the subject of the British physiologist Augustus Waller's (1856–1922) studies. He conducted experiments on nerve fibers in frogs' tongues, and was the first to describe a nerve's appearance when the axon was cut (in other words, when it became a degenerative nerve). From this research, Waller concluded that the cell body is the axon's source of life, and he established that the importance of the nucleus is nerve fiber regeneration.

NEUROSCIENCE ADVANCEMENTS

The second half of the nineteenth century saw great advancements in the study of the nerves and neuroanatomy.

In the 1860s, research by Otto Friedrich Karl Deiters (1834–1863) proposed that neuron cells had two different kinds of branched mechanisms, which eventually became defined as dendrites and axons. These cells were defined as "neurons" in 1891 by Wilhelm van Waldeyer (1836–1921). In 1870, the German physiologist Julius Bernstein (1839–1917) published explanations of nerve membrane polarization. He believed that the membrane's inactive nerve or muscle fiber is electrically polarized and that the external surface is positive in relation to the internal one. The action potential, then, is a self-propagated depolarization of the membrane.

BRAIN LOCALIZATION

The mid-nineteenth century also saw the advancement in understanding how the brain operates and functions. An Austrian anatomist and physiologist, Franz Joseph Gall (1758–1828), determined that the brain's gray matter contained active tissue (or neurons), whereas the white matter contained conducting tissue (or ganglia). He is also considered the father of phrenology, which became popular in the nineteenth and twentieth centuries even though it was eventually discredited by scientific research. Phrenologists attempted to determine one's intellect and personality based on the shape of one's skull. They believed that because mental functions are localized, or located, in specific regions of the brain (and human behavior is subsequently dependent on these functions), the skull's surface must reflect the development of these specific regions. But Gall's leadership in the phrenology field angered the Austrian government. In 1802, the Austrian government declared that phrenology was contrary to religious beliefs, and Gall was forced to leave the country in 1805.

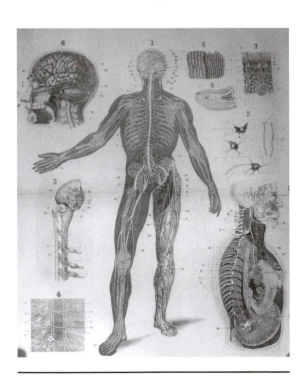

An 1857 depiction of the nervous system, from the *Atlas of Human Anatomy and Physiology*, by William Turner and John Goodsir. © National Library of Medicine.

In the early 1800s, a French physiologist by the name of Marie-Jean-Pierre Flourens (1794–1867) became the first person to demonstrate the brain's function according to its major sections or cortexes. During 1814–1822, Flourens analyzed the physiological changes in pigeons by removing portions of their brains. Throughout these experiments, he determined that by removing the brain's cerebral hemispheres, the functions of will, judgment, and perception were destroyed. He found that if he removed the cerebellum, the pigeons lost muscular coordination and equilibrium. He also learned that death would result if the back portions of the brain, the medulla oblongata, were removed. Based on these findings, Flourens concluded that the cerebral portions of the brain control intellectual functions; the cerebellum controls movement and motor activity; and the medulla controls vital functions, such as breathing. In addition, he determined that equilibrium and coordination were maintained by semicircular canals in the inner ear.

Not surprisingly, Flourens argued against phrenology. He also did extensive research into the effect of lesions on the brain, and determined that lesions did lead to brain impairment. But he also believed that if the lesions were not too severe, some functional recovery could occur and some functional reestablishment was possible because the remaining areas of the brain tended to compensate for the injured portion.

Further proving Gall and Flourens's brain localization beliefs, the French surgeon Paul Broca (1824–1880) demonstrated in 1861 that the speech function was located in the brain (this region of the brain was named **Broca's area**; see Chapter 3). Broca was also a phrenology critic, and he overthrew Gall and the phrenology theories by showing that because the thickness of the skull varies in different parts of the head, it certainly doesn't reflect the brain's true topography.

Another trailblazing doctor around the turn of the century was a German by the name of Carl Wernicke (1848–1904), who also focused his research on nerve disorders involving speech or the ability to communicate, which are known as **aphasias**. In addition, he discovered that the receptive language area of the brain was located in the posterior one third of the superior temporal cortex. Beginning in 1874, Wernicke published papers defining sensory aphasia and its relation to the brain. His 1881 book, *Text-*

Thy puny arm, with principles contend!
No! rather make of every truth a friend;
Take to thyself thy spirit's richest dower,
And arm thy being with its innate power;
Then shalt thou conquer by the might of right,
And pour upon the world a blazing light.

Franz Joseph Gall holding a model of the human head. Wood engraving. © National Library of Medicine.

book of Brain Disorders, attempted to account for the cerebral localization of all known neurologic diseases.

TURN OF THE CENTURY BRINGS MORE KNOWLEDGE OF THE BRAIN

With the advancements made in science throughout the eighteenth and nineteenth centuries, it's not surprising that medicine, in particular surgery, was revolutionized by the start of the twentieth century. With the development of anesthesia in the mid-1850s, comprehensive surgery became safer and much more common. By the beginning of the twentieth century, surgery was used to treat gastrointestinal ulcers and tumors. In fact, many historians call the twentieth century the "golden age of surgery."

But surgery of the nervous system was almost entirely a twentieth-century advancement. The first neurosurgical specialist is considered to be Harvey Cushing (1869–1939). In 1912, Cushing became Harvard University's surgical professor and published his infamous work entitled *The Pituitary Body and Its Disorders*. Through his surgical practice, Cushing became the first person to electrically stimulate the human sensory cortex of the brain.

NOBEL PRIZES MARK MODERN NEUROSCIENCE ACHIEVEMENTS

Major discoveries of the brain and nervous system in the twentieth and twenty-first centuries were recognized by the Nobel Prize, which celebrates not only international scientific achievements but also advancements in physics, chemistry, medicine or physiology, literature, and peace. The Nobel Prize is named after Alfred Nobel (1833–1895), a Swedish scientist who invented dynamite in 1866, in addition to holding more than 350 patents throughout his life. In his will, he declared that money from his estate should be used to establish a private institution called the Nobel Foundation, which governs the Nobel Prize. The Foundation was established in 1900, and the first prize was awarded in 1901. Chapter 7 features a detailed chronology and description of neuroscience advancements that were awarded the Nobel Prize in medicine or physiology.

Nobel Prize Winners in Medicine and Physiology Pursue Neuroscience Discoveries

Born in 1833 in Stockholm, Sweden, Alfred Nobel studied to be an engineer, but later went on to invent dynamite in 1866. Throughout his life, Nobel's intellectual pursuits involved literature and promoting peace in the political arena. But he was also passionate about scientific knowledge, which enabled him to invent dynamite, among other discoveries for which he held over 350 patents. Before dying of a brain hemorrhage in 1895, Nobel stated in his will that a portion of his estate should go toward establishing a foundation that will annually distribute prizes to those "who, during the preceding year, shall have conferred the greatest benefit on mankind." He went on to outline categories for the prizes—physics, chemistry, literature, medicine or physiology, and peace. In 1969, the Nobel Foundation added another category, economics, to the prize distribution.

This chapter will highlight those winners, or "laureates," whose discoveries of the nervous system were recognized in the medicine or physiology category (each winner or winners will be listed following the year they were honored). Since the first prize was given out in 1901, there have been 172 laureates in this category, although many of these involved the heart. But when Professor G. Liljestrand, a member of Stockholm's Royal Caroline Institute, presented the Nobel Prize to Sir Charles Scott Sherrington and Edgar Douglas Adrian in 1932, he emphasized the importance of the advancements in the field of neuroscience:

Within the domain of physiology and medicine probably few spheres will be calculated to attract to themselves attention to the same extent as the nervous system, that distributor of rapid messages between the various parts of the body, and beyond that the material foundation of mental life. . . . To obtain a clearer insight into this complicated machinery, its construction, and its own proper features, has been associated with great difficulties.

Presentation Speech, Liljestrand, 1932

For a comprehensive listing of winners from all categories, visit the Nobel e-Museum on the Internet at www.nobel.se. Table 7.1 features a brief description of all the neuroscience laureates detailed below.

1906—CAMILLO GOLGI (1843–1926) AND SANTIAGO RAMÓN Y CAJAL (1852–1934)

Nobel Prize winner Santiago Ramón y Cajal. © National Library of Medicine.

An Italian physician, Golgi, and a Spanish doctor with a love for drawing anatomy, Ramón y Cajal, were honored with the Nobel Prize for their investigations into the fine, detailed structure of the nervous system. In 1873, Golgi developed a way of staining individual nerve and cell structures, which was later known as the black reaction. By using a weak solution of silver nitrate, Golgi was able to demonstrate the existence of a specific kind of nerve cell, which was subsequently called the Golgi cell. With this staining method, Golgi was able to show intricate details of the nerve cell, including the branching dendrites, or extensions, that connected to other nerve cells.

Ramón y Cajal (see photo) improved on Golgi's staining technique by developing a gold stain in 1913, which allowed scientists to view and study the nervous tissue in the brain and sensory centers, in addition to the spinal cords of embryos and young animals. These staining techniques enabled Ramón y Cajal to distinguish neurons from other cells and to trace the structure of nerve cells and how they connect to the brain's

TABLE 7.1. Nobel Prize Winners in Neuroscience

Year	Name	Research
1906	Camillo Golgi (1843–1926) and Santiago Ramón y Cajal (1852–1934)	Detailed structure of the nervous system
1911	Allvar Gullstrand (1862–1930)	Optical operations of the eye
1914	Robert Bárány (1876–1936)	Research on the vestibular apparatus, equilibrium and balance mechanisms
1927	Julius Wagner–Jauregg (1857–1940)	Malaria inoculation to treat dementia
1932	Sir Charles Scott Sherrington (1857–1952) and Edgar Douglas Adrian (1889–1977)	Nerve cell functions in the brain and spinal cord, including sending messages to muscles
1936	Sir Henry Hallett Dale (1875–1968) and Otto Loewi (1873–1961)	Chemical transmission of nerve impulses and the function of neurotransmitters
1944	Joseph Erlanger (1874–1965) and Herbert Spencer Gasser (1888–1963)	Electrical response of nerve fibers
1949	António Egas Moniz (1847–1955) and Walter Rudolph Hess (1881–1973)	Research on mental disorders, including the development of the prefrontal lobotomy
1957	Daniel Bovet (1907–1992)	Development of various chemotherapeutic agents, or muscle relaxants
1961	Georg von Békésy (1899–1972)	Discovery of the physical means by which sound is analyzed and communicated in the cochlea, a section of the inner ear
1963	John Carew Eccles (1903–1997), Alan Hodgkin (1914–1998), and Andrew Fielding Huxley (b. 1917)	Chemical processes involved when impulses pass along individual nerve fibers
1967	Ragnar Arthur Granit (1900–1991), Halden Keffer Hartline (1903–1983), and George Wald (1906–1997)	Chemistry and physiology of vision
1970	Julius Axelrod (b. 1912), Bernard Katz (1911–2003), and Ulf von Euler (1905–1983)	Electrical operations of synapses, neurotransmitters, and nerve fibers
1973	Karl von Frisch (1886–1982), Konrad Lorenz (1903–1989), and Nikolaas Tinbergen (1907–1988)	Study of animal behavior patterns and ethology, which is observing animals in their natural environments
1976	Baruch S. Blumberg (b. 1925) and Daniel C. Gajdusek (b. 1923)	Research on the origins and behavior of viral infectious diseases, such as hepatitis B
1977	Roger Guillemin (b. 1924) and Andrew V. Schally (b. 1926)	Research on hormones released by the hypothalamus that regulate the pituitary gland

TABLE 7.1. (continued)

Year	Name	Research
1979	Godfrey N. Hounsfield (b. 1919) and Allan M. Cormack (1924–1998)	Development of the computerized axial tomography (CAT) technique used for viewing body tissues through x-ray imaging
1981	David H. Hubel (b. 1926), Roger W. Sperry (1913–1994), and Torsten N. Wiesel (b. 1924)	Studies of the brain functions, in particular how the brain's visual cortex processes information
1982	Sune K. Bergström (b. 1916), Bengt I. Samuelsson (b. 1934), and John R. Vane (b. 1927)	Analysis of prostaglandins, a related group of naturally occurring compounds that influences many physiological functions, such as blood pressure, body temperature, and allergic reactions
1986	Stanley Cohen (b. 1922) and Rita Levi-Montalcini (b. 1909)	Research on the growth of nerve cells and fibers
1991	Erwin Neher (b. 1944) and Bert Sakmann (b. 1942)	Research on basic cell function in addition to the development of the patch-clamp
1994	Alfred Gilman (b. 1941) and Martin Rodbell (1925–1998)	Research on how cells communicate
1997	Stanley Prusiner (b. 1942)	Discovery of prions, disease-causing germs linked to brain-damaging diseases
2000	Arvid Carlsson (b. 1923), Paul Greengard (b. 1925), and Eric Kandel (b. 1929)	Research on how brain cells transmit signals to each other, which led to significant advances in treating diseases afflicting the nervous system

gray matter and the spinal cord. This discovery allowed Ramón y Cajal and his colleagues to declare that the nerve cell is the basic structural unit of the nervous system, which was a critical landmark in the study of neurology and neuroscience.

1911—ALLVAR GULLSTRAND (1862–1930)

This Swedish doctor focused his research on the eye and the optical functions of the nervous system, distinguishing himself early in his career as a renowned ophthalmologist, or eye doctor. Gullstrand was able to determine exactly how optical images are formed in the eye, which included understanding refraction, or how light bends in the eye to produce an image. Gullstrand also determined how the eye's lens changes shape in order to clearly see images at various distances. By contributing to the knowledge of the structure and the function of the eye's cornea, Gullstrand's research enabled scientists to understand the eye's physical operation in order to develop an understanding for how it operates as a sensory organ.

1914—ROBERT BÁRÁNY (1876–1936)

A native of Austria, Bárány was honored for his research on the vestibular apparatus, which is the inner ear's chamber that contains the utricle and saccule—both vital to maintaining equilibrium and balance. Bárány determined that balance was controlled by vestibular reaction movements and involved the positioning of the head and the muscles in the trunk of the body. In addition, he researched the connection between the vestibular apparatus and the corresponding areas of the brain, or the cerebellum, that process vibrations, produce sounds, and establish equilibrium.

1927—JULIUS WAGNER-JAUREGG (1857–1940)

Since the time of Hippocrates, scientists and doctors had noted that patients that had a paralyzing form of chronic mental illness (referred to as "dementia paralytica" in 1927 by the Nobel committee) were cured or treated successfully when they were attacked by a fever-inducing disease. Wagner-Jauregg tested this theory by injecting nine patients suffering from mental illness with the infectious blood from malaria patients. All the patients reacted favorably to the treatment, and three were even cured (also, it's important to note that the form of malaria he injected the patients with was fairly innocuous and could be easily cured with quinine). This artificial introduction of malaria constituted the first example of shock therapy. Prior to Wagner-Jauregg's research, the mentally ill were considered beyond help and incurable. However, Wagner-Jauregg's studies marked the beginning of research aimed at curing or treating those patients with chronic mental illness. His malaria treatment was later replaced with the administration of antibiotics, but Wagner-Jauregg's theoretical research led to the development of fever and shock therapy for a number of mental disorders.

1932—SIR CHARLES SCOTT SHERRINGTON (1857–1952) AND EDGAR DOUGLAS ADRIAN (1889–1977)

Almost thirty years after Golgi and Ramón y Cajal established that neurons were the foundation of the nervous system, another partnership, formed by the British scientists Sherrington and Adrian, was honored in 1932 for their discoveries regarding the nerve cell. Sherrington's research focused on animals, such as cats, dogs, and monkeys, whose brains had been deprived of certain hemispheres. He found that reflexes are a result of integrated (rather than isolated) activity of the entire nervous system. Out of this research came **Sherrington's Law**, which states that when one set of muscles is stimulated, the opposite muscles, or those opposing the action, are inhibited. In addition, Sherrington coined the term "synapse" to distinguish the nerve cell and the point at which a nervous impulse is transmitted from one nerve cell to another.

Adrian was fascinated with the sense organs that received and transmitted these stimuli. He studied how the body reacted to intense stimuli, and found that nerve fibers exerted the same force, regardless of the power or intensity of the stimuli. What varied was the speed at which the neuron fiber transmitted the data to the motor neuron—stronger stimuli elicited a faster response, because it had excited a greater number of nerve fibers in comparison to a weaker stimulus. Therefore, although the signals or reactions are the same throughout the body, the reaction depends on how many nerve fibers are stimulated or excited. Later in his research, Adrian recorded data from nerve impulses from single-sensory endings and then the reacting motor nerve fibers. These measurements enabled a clearer understanding of the physical expressions of sensation and how muscular control operates. After he won the Nobel Prize, Adrian studied the brain's electrical activity as it relates to abnormalities, such as cerebral lesions and disorders such as epilepsy.

1936—SIR HENRY HALLETT DALE (1875–1968) AND OTTO LOEWI (1873–1961)

Both Dale and Loewi were honored with the Nobel Prize in 1936 for their research of the chemical transmission of nerve impulses. Loewi's research provided the first proof that chemicals were involved in helping nerve impulses cross synapses en route to the responsive organ.

Dale published extensive investigations, the most famous of which involved isolating acetycholine in animal tissue. After researching this chemical in a frog's heart, he proved that acetylcholine is the neurotransmitter that helps nerve impulses to cross synapses (see Chapter 4) in both the sympathetic and parasympathetic divisions of the autonomic nervous system.

1944—JOSEPH ERLANGER (1874–1965) AND HERBERT SPENCER GASSER (1888–1963)

This American partnership began in 1906, when Gasser was one of Erlanger's students at the University of Wisconsin–Madison. Gasser followed Erlanger to his next position at Washington University in St. Louis, Missouri, where the duo began researching the electrical responses of nerve fibers. They were able to isolate and amplify the electrical wave of an impulse generated in a single nerve fiber. Once this was amplified, the various components of the nerve's response could be studied.

In 1932, Erlanger and Gasser determined that nerve fibers conduct impulses at different rates, depending on the thickness of the fiber. In addition, they proved that each fiber has a different threshold of stimulation, meaning that each fiber requires a stimulus of a different intensity to generate an impulse. Based on this discovery, they demonstrated that different nerve fibers conducted specific kinds of impulses, such as pain, cold, or heat. Erlanger and Gasser's work led to the implementation of recording machines to capture and amplify these impulse waves in order to diagnose and treat brain and nervous system disorders.

1949—ANTÓNIO EGAS MONIZ (1847–1955) AND WALTER RUDOLPH HESS (1881–1973)

Moniz, who is known as the founder of modern psychosurgery, and Hess were awarded the Nobel Prize in 1949 for the development of the **prefrontal lobotomy**, a radical brain surgery for certain brain disorders.

As a neurology professor at Portugal's University of Lisbon, Moniz developed a method called **cerebral angiography**. By injecting certain substances into the brain's blood vessels, he was able to view the brain and its components more clearly on an x-ray exam. Through this testing, Moniz noticed that certain mental disorders, such as **schizophrenia** and **paranoia** (which will be further explored in Chapter 8), involved recurrent thought patterns that dominated one's psychology. Therefore, he believed that by severing the nerve fibers present in the brain's frontal lobes (which are responsible for psychological responses) from the thalamus (a relay center for the brain's sensory impulses), thought processes might be normalized. This operation was known as a prefrontal lobotomy and is now known to have serious side effects. Although lobotomies were somewhat widespread in the 1940s and 1950s, today pharmaceutical remedies are relied on for treatment instead.

Whereas Moniz's research centered on the brain, Hess's work focused on the autonomic nervous system, whose nerves originate at the base of the brain and extend throughout the spinal cord and control such functions as digestion. Hess's research focused on cats and dogs. By electrically stimulating specific areas of their brains, he was able to determine that the body's autonomous functions are centered at the brain's base, in the medulla oblongata and the hypothalamus.

1957—DANIEL BOVET (1907–1992)

This Swiss-born pharmacologist is responsible for the development of various chemotherapeutic agents, or muscle relaxants. His discoveries of substances such as gallamine were eventually used in conjunction with anesthesia to induce muscle relaxation during surgery.

1961—GEORG VON BÉKÉSY (1899–1972)

This American scientist was awarded the Nobel Prize for his discovery of the physical means by which sound is analyzed and communicated in the cochlea, a section of the inner ear. Although it had been known since the mid-nineteenth century that the basilar membrane is the vibratory tissue responsible for hearing, Békésy determined that sound travels along this membrane in a series of waves that peak at different places. For instance, low frequencies peak toward the end of the cochlea and high frequencies peak near its base, or entrance. The location of the nerve receptors (in addition to the number of receptors involved) determines pitch and loudness of sound.

1963—JOHN CAREW ECCLES (1903–1997), ALAN HODGKIN (1914–1998), AND ANDREW FIELDING HUXLEY (b. 1917)

This trio furthered Dale and Loewi's work studying the chemical processes involved when impulses pass along individual nerve fibers.

In the early 1950s, Hodgkin and Huxley worked together at Cambridge University researching the electrical and chemical behavior of individual nerve fibers. By examining the giant nerve fibers of a squid, the team was able to demonstrate that the electrical potential of a fiber during conduction is greater than when the nerve is resting. This theory replaced previously held beliefs that the nerve membrane broke down during impulse conduction.

Hodgkin and Huxley also believed that the activity of a nerve fiber depends largely on the fact that a large concentration of potassium ions is located inside the fiber and a large concentration of sodium ions is found outside of the fiber. The exchange of sodium and potassium ions causes a temporary reversal of the nerve cell's action potential, or electrical polarization. Building on this research, Eccles showed that when a nerve cell is excited, one kind of synapse releases a substance such as the neurotransmitter acetylcholine. This release causes the pores in the nerve membrane to expand, which allows sodium ions to flow inside of the nerve cell, thus reversing the polarity of the electrical charge. This constitutes a nerve impulse. In the same way, however, Eccles showed that the excited nerve cells prompt the synapse to release another kind of substance that forces the sodium ions to pass outside of the cell, thus inhibiting the impulse transmission (see Chapter 1).

1967—RAGNAR ARTHUR GRANIT (1900–1991), HALDEN KEFFER HARTLINE (1903–1983), AND GEORGE WALD (1906–1997)

Granit, Hartline, and Wald were awarded the Nobel Prize in 1967 for their research on the chemistry and physiology of vision.

Granit is famous for his "dominator-modulator" theory of color vision. He knew that the eye had three kinds of photosensitive cones, which are the color receptors located in the retina that responded to different portions of the light spectrum. But he also believed that some optic nerve fibers, which he called dominators, are sensitive to the entire light spectrum, and that others, which he named modulators, only respond to a narrow band and are largely responsible for color vision.

Hartline focused his research on the electrical responses of the retina. Focusing on animals with simple visual systems, such as mollusks, he demonstrated the first record of electrical impulses sent by a single optic nerve fiber when the receptors are stimulated by light. By examining this research, Hartline determined that the eyes' receptor cells are interconnected so that whereas one cell might be stimulated by light, other cells located nearby are depressed. This contributes to the contrast in light patterns and sharpens the perception of shapes. Hartline continued researching how various mechanisms of the retina contribute to the integration of visual information.

Wald is most famous for his discovery of the importance of vitamin A in preserving the retina and thus maintaining vision. In the late 1950s, he noted that certain pigments in the retina were sensitive to yellow-green and red light. A decade later, he discovered the pigment sensitive to blue light, in addition to the role that vitamin A played in forming these color pigments and the fact that color blindness is caused by the absence of one of these pigments.

1970—JULIUS AXELROD (b. 1912), BERNARD KATZ (1911–2003), AND ULF VON EULER (1905–1983)

These three scientists were awarded the Nobel Prize in 1970 for their work on neurotransmitters and nerve fibers.

Katz published three landmark books—*Electric Excitation of Nerve* (1939), *Nerve, Muscle and Synapse* (1966), and *The Release of Neural Transmitter Substances* (1969)—which focused on the chemical aspects of nerve transmission. Katz's work also included the discovery of the role that calcium ions had in promoting the release of neurotransmitters.

Von Euler identified one of these neurotransmitters, norepinephrine, which is the most important impulse carrier in the sympathetic nervous system. He also discovered hormones called prostaglandins, which are key in stimulating muscular contraction in the cardiovascular and nervous systems.

Building upon von Euler's identification of norepinephrine, Axelrod discovered that this neurotransmitter could be inactivated by an enzyme called catechol-O-methyl transferase. This enzyme proved vital in the creation of certain drugs for mental illness, such as schizophrenia, in addition to hypertension.

1973—KARL VON FRISCH (1886–1982), KONRAD LORENZ (1903–1989), AND NIKOLAAS TINBERGEN (1907–1988)

In 1973, von Frisch, Lorenz, and Tinbergen were awarded the Nobel Prize for their work on animal behavior patterns. All three are credited with reviving interest in the study of **ethology**, which is observing animals in their natural environments.

Tinbergen's research focused on how both instinctive and learned behavior help an animal survive. Based on these observations, he examined the nature of human violence and aggression. He is well known for his observations of sea gulls, in which he noted their courtship and mating behavior.

Von Frisch (see photo) was recognized for his work with bees. He demonstrated that bees could be trained to distinguish various tastes and smells, and also found that bees communicate with each other regarding food supply by certain types of rhythmic dancing, such as wagging and circling.

Lorenz began his animal research following young ducklings and goslings. He learned that soon after hatching, these animals learned to follow their parent figure and imitated both visual and auditory behavior, which is known as **imprinting**. For example, young ducklings learn to quack and make similar noises by imitating the adults around them. Lorenz also worked with Tinbergen to study the nature of instinctive behavior, in particular how some violence and aggressive acts come about and what constitutes the accompanying nervous energy. In addition, Lorenz spent a considerable amount of time studying long-term behavioral patterns of species, and theorized that organisms are genetically constructed to adapt and learn specific information that contributes to the survival of the species.

Nobel Prize winner Karl von Frisch in a botanical collection room. © National Library of Medicine.

1976—BARUCH S. BLUMBERG (b. 1925) AND DANIEL C. GAJDUSEK (b. 1923)

Blumberg and Gajdusek were presented the Nobel Prize in 1976 for their research on the origins and behavior of viral infectious diseases, such as hepatitis B. In 1967, Blumberg identified an **antigen** (a foreign substance such as bacteria or a virus that causes the immune system to produce antibodies) in the blood of an Australian aborigine that was part of a virus that causes hepatitis B. The antigen proved to protect the body against infection from the virus that caused the disease, which eventually led to a safe and effective vaccine against the disease made commercially available in the United States in 1982.

Gajdusek also did significant research in Australia, where he studied a degenerative brain disorder among a group called the Fore people in New Guinea. Through living among these people and studying their culture and language, in addition to performing autopsies on their dead, he found that the disease was transmitted when the Fore people ate the brains of the deceased members of the group, a ritual that was an intrinsic part of the funeral procession. Because the disease had a delayed onset, or didn't appear to show symptoms until midlife, Gajdusek theorized the disease was caused by a virus that had the ability to lay dormant in the body for years. His work had significant impact on future research on degenerative brain disorders.

1977—ROGER GUILLEMIN (b. 1924) AND ANDREW V. SCHALLY (b. 1926)

Guillemin and Schally were awarded the Nobel Prize in 1977 for their research on hormones released by the hypothalamus that regulate the pituitary gland. Guillemin discovered, isolated, and analyzed TRH, which is the thyrotropin-releasing hormone that regulates thyroid activity; GHRH, which is the growth hormone–releasing hormone; and somatostatin, which regulates the pancreas in addition to the pituitary gland. In addition, he discovered **endorphins**, which are proteins involved in the body and mind's perception of pain. Schally also researched TRH, in addition to isolating the LH-RH (luteinizing hormone–releasing hormone).

1979—GODFREY N. HOUNSFIELD (b. 1919) AND ALLAN M. CORMACK (1924–1998)

These two scientists were recognized for their work in developing the **computerized axial tomography (CAT)** technique used for viewing body tissues through x-ray imaging.

After studying electronics and radar while in the Royal Air Force during World War II, Hounsfield led a team of scientists and researchers to create one of the first computers in 1958–1959. Combining the capability of the computer with x-ray technology, Hounsfield was able to produce a two-dimensional image of the human head. He used a technique called axial tomography that enabled him to build a head scanner. Cormack was vital to this discovery, because as a trained physicist, he developed the mathematical technique eventually used by the computers to calculate the various measurements from different angles and dimensions. Computers were eventually able to process data from the scanner, enabling the first successful CAT scan to be performed in 1972.

1981—DAVID H. HUBEL (b. 1926), ROGER W. SPERRY (1913–1994), AND TORSTEN N. WIESEL (b. 1924)

This trio of scientists was honored for their studies of the brain functions, in particular how the brain's visual cortex processes information. Sperry focused his research on animals and human brains in which the nerve network out of the corpus callosum that connects the right and left hemispheres had been severed. Through examining these brains, he found that the left side of the brain or hemisphere is usually analytically and verbally dominant, whereas the right side or hemisphere in dominant is creative tasks such as music. Sperry developed surgical and experimental techniques that allowed future researchers to explore the brain's specialized functions and how they relate to different areas of the brain.

Hubel and Wiesel began working together at Harvard Medical School in 1959 researching the path of nerve impulses from the retina to the brain's sensory and motor areas. The team measured and tracked the electrical discharges occurring in the individual nerve fibers and brain cells in the visual cortex when the retina responded to light. They also looked at how data are processed and passed along to the brain. Wiesel and Hubel also demonstrated the importance of treating eye defects that are detectable in young and even newborn children.

1982—SUNE K. BERGSTRÖM (b. 1916), BENGT I. SAMUELSSON (b. 1934), AND JOHN R. VANE (b. 1927)

This trio of researchers was honored with a Nobel Prize in 1982 for their analysis of prostaglandins. This related group of naturally occurring compounds influences many physiological functions, such as blood pressure, body temperature, and allergic reactions.

Samuelsson and Bergström joined together in the early 1960s, and became the first to identify the molecular structure of a prostaglandin. Through their research, they determined that these compounds are derived from oxygen that combines with arachidonic acid, a type of fatty acid found in various kinds of meat and vegetable oil. Samuelsson eventually showed how certain prostaglandins are involved in blood clotting and the contraction of blood vessels. In his work with prostaglandins, Vane demonstrated how aspirin works and why it has become the world's most widely used drug, because it inhibits the development of prostaglandins associated with pain, fever, and inflammation.

1986—STANLEY COHEN (b. 1922) AND RITA LEVI-MONTALCINI (b. 1909)

The Brooklyn-born biochemist and Italian female neurologist shared the 1986 Nobel Prize for their research on nerve tissues and cells.

Because of her Jewish ancestry, Levi-Montalcini went into hiding while the Germans occupied Italy during World War II. When the war ended, she came to the United States, where she studied the growth of nerve tissues in chick embryos. Through various experiments with mice and these embryos, she discovered that implanting a certain type of mouse tumor into a chick embryo encouraged nerve growth. Levi-Montalcini and her colleagues found the cause was a specific substance in the tumor that they named the **nerve-growth factor (NGF)**.

Cohen then joined Levi-Montalcini's research team and found another substance, the **epidermal growth factor (EGF)**, which that caused the eyes and teeth of newborn laboratory mice to develop sooner than normal. EGF was later found to play an important role of several development events in the human body, and NGF was determined to play an important role in the growth of nerve cells and fibers in the nervous system.

1991—ERWIN NEHER (b. 1944) AND BERT SAKMANN (b. 1942)

Neher and Sakmann were awarded the Nobel Prize in 1991 for their basic cell function research in addition to the 1976 development of a technique called the patch-clamp, which can detect small electrical currents produced by ions traveling through cell membranes. Composed of many porous channels, the cell membrane directs the passage of ions, or charged atoms, as they pass into and out of the cell. Neher and Sakmann paired a thin glass instrument called a pipette and an **electrode** to detect the flow of ions through the cell membrane. Through the patch-clamp, doctors were able to eventually determine the role that ions played in diseases such as **diabetes**, **cystic fibrosis**, **epilepsy**, and several other cardiovascular disorders. This invention also furthered research into drugs and medications targeted towards these diseases.

1994—ALFRED GILMAN (b. 1941) AND MARTIN RODBELL (1925–1998)

Gilman and Rodbell won the Nobel Prize in 1994 for their separate research of certain molecules that cells use to process an incoming signal, such as a hormone or neurotransmitter. Before this duo's work, researchers and scientists believed that cells communicated using a hormone receptor and an interior cell enzyme that amplifies the signal. But Rodbell discovered that there was actually an intermediary communicator between the receptor and the enzyme to relay the message between the receiver and amplifier. Gilman identified this signaling communicator as the **G-protein**, which is named because it is activated when bound to a molecule called guanosine triphosphate (GTP). After Gilman and Rodbell's discovery, over twenty G-proteins were identified, which led to a better understanding of diseases such as cholera, diabetes, alcoholism, and cancer.

1997—STANLEY PRUSINER (b. 1942)

Prusiner was awarded the 1997 Nobel Prize for discovering **prions**, disease-causing germs linked to brain-damaging diseases. This American biologist began his research in 1972 studying the Cruetzfeldt-Jacob disease, the human form of "mad cow" disease. This led to an increased understanding of mad cow, or bovine spongiform encephalopathy, along with Alzheimer's disease and other illnesses that cause **dementia**.

2000—ARVID CARLSSON (b. 1923), PAUL GREENGARD (b. 1925), AND ERIC KANDEL (b. 1929)

These three scientists were awarded the Nobel Prize in 2000 for research on how brain cells transmit signals to each other, which led to significant advances in treating diseases afflicting the nervous system. Carlsson, in particular, was singled out for his discovery of a drug used to treat Parkinson's disease, and Greengard was recognized for his research on how the brain transmits data to the nervous system. Kandel was recognized for his studies on long- and short-term memory.

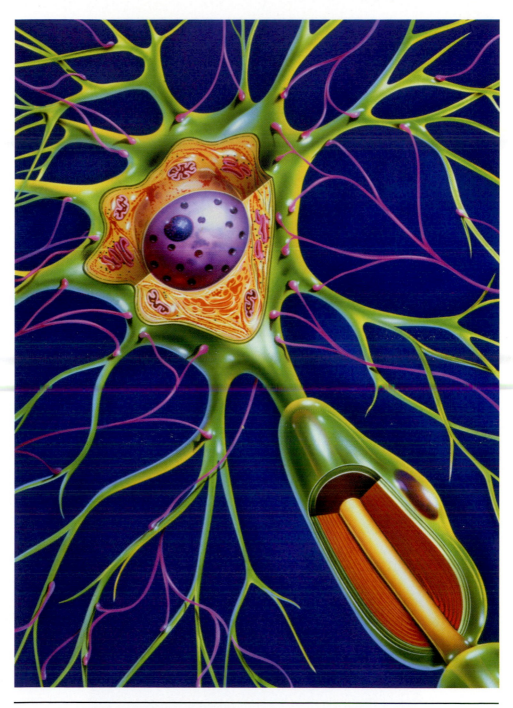

Nerve cell or neuron. © JOHN BAVOSI/SPL/Custom Medical Stock Photo.

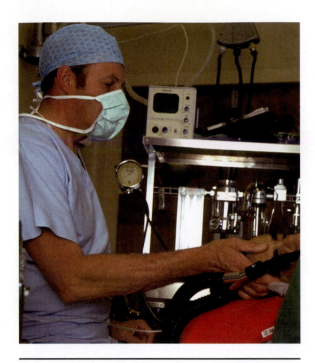

An anesthetist in an operating theater. © Mediscan/Visuals Unlimited.

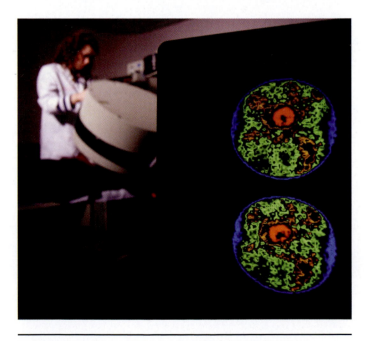

Doctor performing a PET scan. © Ed Eckstein/Phototake.

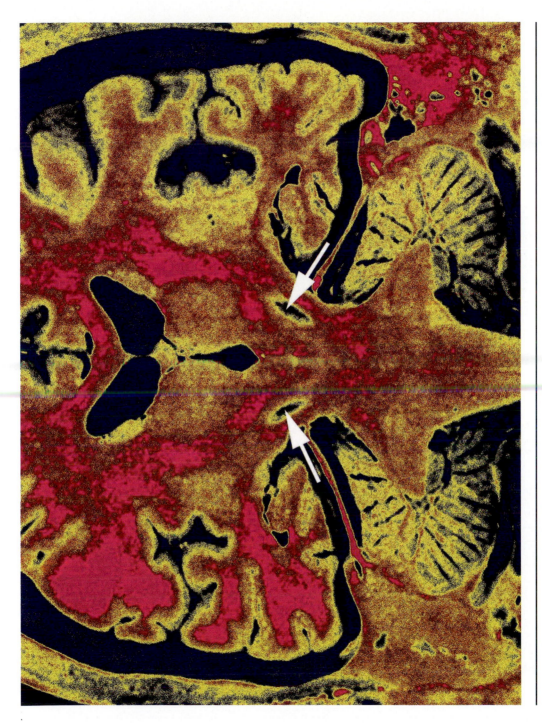

Parkinson's brain disease. © J. Cavallini/Custom Medical Stock Photo.

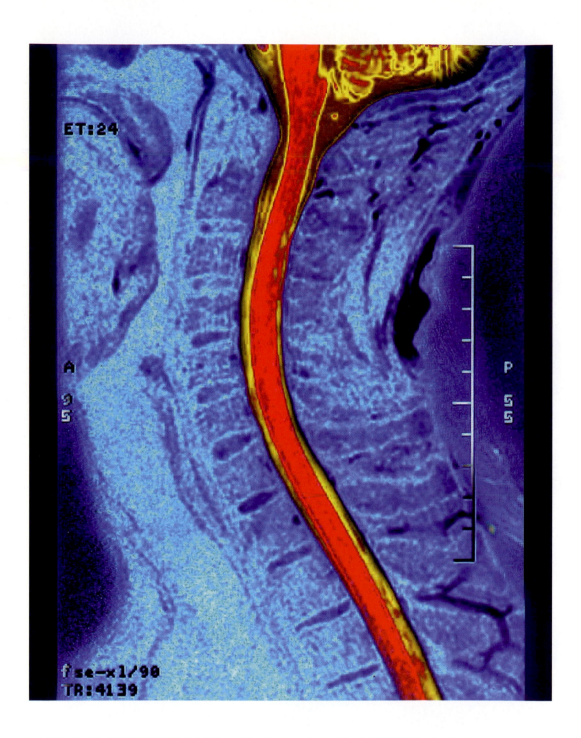

The medulla spinalis or spinal cord. © J. Cavallini/Custom Medical Stock Photo.

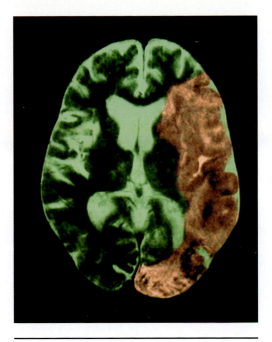

An MRI of a massive stroke in a human brain. ©
Howard J. Radzyner/Phototake.

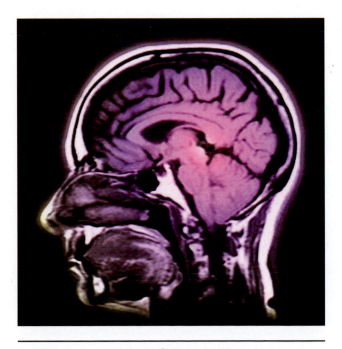

An MRI of a woman's head. © Photodisc/Getty Images.

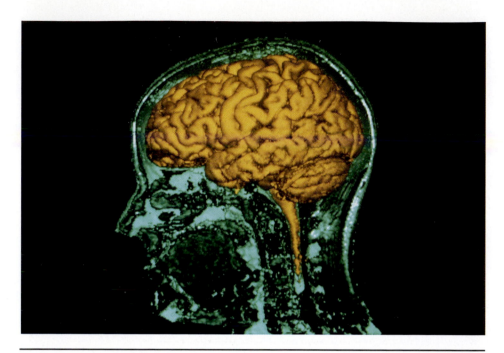

3D NMR (nuclear magnetic resonance) scan of a normal brain. © Collection CNRI/Phototake.

A retired 85-year-old in his workshop. © Skjold Photographs.

Color engraving of René Descartes. © National Library of Medicine.

Portrait of Christopher Reeve. Courtesy, Christopher Reeve Paralysis Foundation.

8

Nervous System Diseases and Disorders

Disorders of the nervous system involve some injury or illness centered in the brain or the spinal cord, or both, including both organs' nerve systems. It's fair to say that nearly every disease and sickness affecting the human body influences the nervous system in some form. This chapter, however, will focus on some of the more common disorders. In some cases, the descriptions will include information on appropriate therapies or medications used to treat the condition.

SPINAL CORD INJURY

A network of nerve cells and fibers approximately 17 inches long, the human spinal cord begins at the base of the brain and extends to the lower back. As explained in Chapter 2, the spinal cord is the body's messenger—receiving data from receptors, taking them to the brain, and then transmitting them to all parts of the body. Encased in the backbone, the spinal cord is made up of thirty-three individual vertebrae (see Figure 1.2). When an injury occurs at any level of the spinal cord, the communication between the brain and the body is disrupted, and messages can no longer travel to the damaged area.

The area or level where the spinal cord is damaged will determine which bodily functions are damaged and the extent of the communication breakdown. The higher the injury is located along the backbone, the greater the damage to movement and sensation. For instance, a thoracic, or upper-back

Garrett Burgess, paralyzed from the neck down due to an automobile accident. He is traveling with his father to raise awareness and funding for spinal cord injuries. © AP/Wide World Photos.

injury, might cause **paraplegia**, affecting only the legs and lower parts of the body. But injury to the cervical, or neck area, of the spinal cord could result in paralysis in both the upper and lower portions of the body. This is referred to as **quadriplegia**, meaning a loss of sensation and function in the legs and arms (see photo). Physicians refer to the specifics of a spinal cord injury using the letter and number coordinates associated with the location of the damage. For instance, a C7 injury means that the seventh cervical spinal cord segment is damaged.

Beyond the letter and number classification, spinal cord injuries are characterized as either "complete" or "incomplete." A complete injury results in total loss of movement and sensation below the level of damage, and an incomplete injury does not result in a total movement and sensation loss. This classification can change, however, as a patient recovers, which is why physicians and therapists will work with patients to maximize whatever degree of function remains after a spinal cord injury.

Effect on Bladder and Bowel Control

In addition to a loss of movement related to arms and walking, a spinal cord injury can result in an impairment of the bowel and bladder, the body's excretion systems. The **bowel**, which controls the removal of solid waste through the body's intestinal tract, is governed by both voluntary and involuntary control. The muscle at the end of the bowel is the anal sphincter, which voluntarily relaxes to allow the intestinal and abdominal muscles to involuntarily release the waste. However, this sphincter muscle may be damaged as a result of the injury, although the involuntary functions will continue to operate. Spinal cord injury patients often need to regularly schedule bowel movements in order to avoid accidents or constipation.

Similar to the bowel, the **bladder** is both under voluntary and involuntary control in order for the body to excrete liquid waste and fluids through the urinary tract. Although the male and female urinary systems are somewhat different, both have kidneys to filter waste and fluid materials not needed by the body. The ureters are the two tubes that connect the kidneys to the bladder, where the filtered waste flows to the outside of the body through another tube called the urethra. In the case of a spinal cord injury, many nerve pathways connecting the brain with the bladder are disrupted. Therefore, the patient often does not know when his bladder is full (the voluntary control). When the bladder gets too full, it will involuntarily empty. In order to prevent urine from backing up from the bladder into the kidneys (which could cause a kidney infection), a small tube called a catheter might be inserted into the patient's bladder to help with excretion.

Spasticity and Pressure Sores

When a spinal injury occurs, the body will be in a state of "spinal shock" for weeks or even months, and normal reflexes are no longer present below the area of injury. After this period passes, however, some amount of normal reflexes will return, and some will even become exaggerated if they are no longer controlled by the brain, which results in **spasticity**, or spasms. This is because the injury has disrupted the flow of information about the environment to the brain, and therefore the brain can no longer control the reflex activity that happens in response to environmental changes, causing the reflexes to become exaggerated or spastic.

Because spinal injury patients are often sitting or lying in one position for long periods, they are likely to develop **pressure sores** where the skin tissue has broken down. The body is susceptible to pressure sores at areas that support the weight during long periods of sitting or lying in bed. The longer the body rests on a surface, the longer the bone will compress the skin, reducing the flow of blood to that area. When this blood flow is blocked for too long, the skin begins to break down. Other areas of the body where the bones come close to the outer skin without a thick layer of fatty tissue for insulation are also at risk for developing pressure sores.

TRAUMATIC BRAIN INJURY

Scientists estimate that approximately 2.5–6.5 million Americans have had a **traumatic brain injury (TBI)** or head injury, leaving them with significant mental and behavior disabilities, in addition to communication difficulties. Young males between the ages of 15 and 24, elderly people 75 and older (of both sexes), and children 5 and younger are at the highest risk for TBI. For people under the age of 75, transportation accidents involving automobiles, motorcycles, and bicycles are the main causes of TBIs. For those 75 and older, a majority of the TBIs are caused by falling. Half of all TBI incidents involve the use of alcohol.

TBI is classified as either a **closed head** or a **penetrating head injury**, and the resulting damage can either be focal or diffuse. A closed head injury occurs when the head suddenly strikes an object, but the object does not break through the skull. In comparison, a penetrating head injury occurs when the object breaks through the skull, piercing the brain tissue. Focal damage is confined to one area of the brain, and diffuse damage involves one or more areas of the brain.

Symptoms of a TBI can be mild, moderate, or severe, and depend on the extent of brain damage. Whereas some symptoms are immediate and obvious, others do not appear for days or weeks after the injury. Mild TBI symptoms include headache, confusion, lightheadedness, dizziness, blurred vision, and changes in mood or behavior. In addition, mild TBI patients might have trouble with memory, concentration, attention, or **cognition**, also known as thinking. Patients with moderate or severe TBI might show these same symptoms. However, they may also have a constant headache, repeated vomiting or nausea, slurred speech, numbness in fingers and toes, increased confusion, and **seizures**.

Following is a list explaining several types of TBI:

Concussion. This is the most minor and common type of TBI. Physicians refer to any minor injury to the head or brain as a **concussion**, but it specifically refers to a short loss of consciousness due to a head injury.

Skull fracture. This type of injury occurs when the skull bone cracks or breaks. A depressed **skull fracture** occurs when skull pieces from that break enter the brain tissue, and a penetrating skull fracture occurs when an object (like a bullet) pierces the skull, causing distinct and specific damage to the brain tissue.

Contusion. A **contusion** is when a skull fracture results in bruising the brain tissue, which is caused by the swollen brain tissue mixing with blood released from broken blood vessels. Another form of a contusion is a **contrecoup**, which is when the brain is injured from repeated shaking back and forth. Shaken baby syndrome occurs when a baby is shaken violently, causing the brain to bounce against the skull repeatedly. Severe contrecoup can lead to diffuse axonal injury, which is when there is severe damage to individual nerve cells in the brain, causing a loss of connections among the neurons. This can lead to a total breakdown of communication among the brain's neuron network.

Hematoma. When a major blood vessel in the head is damaged, heavy bleeding into or around the brain can result. There are three types of **hematomas** that can result in brain damage (refer to Figure 3.1 in Chapter 3 to locate the areas on the brain): epidural hematoma, which occurs when there is bleeding into the area between the skull and the dura; subdural hematoma, when bleeding is confined to the area between the brain's dura and arachnoid membrane; and intracerebral hematoma, when there is bleeding within brain itself.

Anoxia. This injury occurs when there is a lack of oxygen to the brain, even if there is an adequate supply of blood (hypoxia refers to a decrease in oxygen supply rather than a complete absence). Brain cells die without oxygen within several minutes. **Anoxia** is often seen in drowning and heart attack victims.

ALZHEIMER'S DISEASE

The progressive loss of intellectual and cognitive abilities is known as dementia, and is often seen as a natural part of aging. However, the most severe form of dementia is Alzheimer's disease (AD), which is not a normal aspect of aging.

The cause of AD remains for the most part a mystery to scientists and researchers. However, it's thought to involve two kinds of neuron deformations in the brain. These deformations were initially called tangles and plaques by the German neurologist Alois Alzheimer (1864–1915) in 1906, who first noted the microscopic changes to the brain caused by the disease. A neurofibrillary tangle occurs when the various components of the inside of a neuron become deformed and clump together. Plaques, however, occur outside the neuron when a particular protein, called a beta amyloid, accumulates in a deposit form in the brain. The brain needs some amount of this protein to function, but an excessive amount inhibits proper brain function. Beta amyloid is similar to cholesterol in this regard (but with much different molecules). The body needs cholesterol to maintain healthy cells, but too much can cause blocked arteries, leading to heart attacks and other problems.

Symptoms

The most obvious and common AD symptom is memory loss, particularly of recent events and familiar information. Initial symptoms can be subtle and mild, such as forgetting someone's name. But as the disease progresses and worsens, more significant dementia occurs, causing a person to become lost in familiar surroundings (such as his own home) or unable to learn new things. Eventually, many AD patients are unable to recognize friends or family members and their personalities change, causing them to become agitated, paranoid, and depressed.

Recent research suggests that the cerebral cortex, which processes visual and spatial information, is damaged in AD brains. In addition, areas of the brain that are important for memory—the basal forebrain and hippocampus—are

Some degree of memory and intellectual ability loss is a normal aspect of aging. © Photodisc/Getty Images.

also thought to be affected by AD. Research also suggests that AD brains have decreased levels of the neurotransmitter acetylcholine.

Treatment

There is currently no cure for AD, and patients usually live 6–8 years following diagnosis. But there are some medications used to treat the disease. One class of drugs is called **cholinesterase inhibitors**, which work to increase the acetylcholine levels by blocking the enzyme cholinesterase, which works to break down acetylcholine. This medication helps slow memory loss, allowing the AD patient to perform daily tasks.

GUILLAIN-BARRÉ SYNDROME

This disorder involves the body's peripheral nervous system, and is extremely rare: It affects only 2 out of 100,000 people per year. An attack begins when **macrophages**, or cells that absorb waste or harmful material, invade a nerve and proceed to erode through the myelin sheath to the axon. Symptoms include tingling or numbness in the limbs, slow nerve reflexes, and general muscle weakness that can lead to respiratory difficulties and heart failure; however, an attack of Guillain-Barré is treatable and rarely fatal. Onset of the disease can occur after a viral or bacterial infection such as a stomach or intestinal illness.

Experts are unsure of the cause, although research has suggested a link with a certain bacterium often found in raw chicken, dirty water, and unpasteurized milk. Symptoms usually occur two to four weeks after a viral or bacterial infection such as a stomach or intestinal virus, infectious mononucleosis, viral hepatitis, or even the common cold. The most common forms of treatment are **plasmapheresis** and high-dose **immunoglobulin** therapy. Plasmapheresis is when a portion of the patient's blood is removed and then the liquid component is cleansed. These clean blood cells are then injected into the body. In high-dose immunoglobulin therapy, small quantities of proteins that the immune system normally uses to attack invading organisms are injected intravenously into the patient at high doses.

MIGRAINES

Far more extreme than "bad headaches," migraines affect 28 million Americans—about 6 percent of men and 18 percent of women, most of whom are between 25 and 55 years of age. The condition is often hereditary—a child has a 50 percent chance of inheriting the propensity to get migraines if one parent suffers and a 75 percent chance if both parents suffer. They typically occur in episodes, and on average, a migraine patient in the United States will suffer one attack per month. Migraines are most prevalent in women during the adult reproductive years, hitting their peak around the age of 40, after which the disorder begins to fade.

Migraines are defined as **vascular headaches**, a name that refers to an abnormal function of the brain's blood vessels or vascular system. Most experts believe that the disorder involves changes in blood flow to the brain. Specifically, migraine patients appear to have blood vessels that overreact to certain triggers. The body's nervous system responds to a trigger by generating spasms in certain nerve-abundant arteries located at the base of the brain. These spasms cause several of these arteries (especially those located in the scalp and neck arteries) that supply blood to the brain to close or constrict, thus reducing the flow of blood to the brain. Simultaneously, the blood-clotting platelets in these arteries release serotonin, which further constricts the arteries. As a result of this reduction in blood flow, the brain is not getting its usual and adequate supply of oxygen, which might prompt neurological symptoms, such as distorted vision and slurred speech. This causes certain arteries in the brain to widen in an attempt to increase the oxygen flow. As the artery dilation spreads to the neck and scalp, prostaglandin production increases, along with the production of other chemicals that cause inflammation and pain sensitivity. These chemicals circulate, and the dilation of the scalp arteries stimulates nociceptors, sensory receptors that respond to pain. At this point, the migraine attack and accompanying pain is well underway.

Migraines are distinct from other headaches, such as tension headaches, because the pain is accompanied by two major groups of symptoms. The first group is gastrointestinal, which means many patients feel nausea, vomit, and get diarrhea during an attack. The second group is neurological and involves an increased sensitization of the senses. Examples of these symptoms include light sensitivity and a vulnerability to loud sounds or strong smells.

Migraines are classified as either aura or nonaura, depending on which symptoms precede an attack. An aura is a distinct set of neurological symptoms that precedes the headache pain. It's most commonly visual, such as bright flashes of light and image distortions. Most migraine sufferers do not get auras, but for those who do, these symptoms usually last about fifteen or twenty minutes. After a brief calming, the symptoms fade away, and the headache takes hold and begins to escalate. Auras are believed to occur when the nervous system is initially responding to a trigger through arterial spasms at the base of the brain.

Migraines can be provoked when there is a fall in estrogen levels as women move between ovulation and menstruation cycles. Decreased estrogen prompts an increase in the production of prostaglandins, a group of fatty acids involved with regulating blood pressure and body temperature, which can cause cranial blood vessels to dilate and become inflamed. In addition, low estrogen levels can lead to a decrease in production of endorphins, the body's natural pain relievers.

Treatment

There is currently no cure for migraines, although many patients take drugs in the **sumatriptan** family, which work to counteract the dilation and inflammation of the brain's blood vessels. This helps to alleviate the pressure on the pain-sensitive nociceptors. In addition, the anticonvulsant drug divalproex and propranolol hydrochloride have been effective in preventing migraines in some patients by halting blood vessel dilation.

TAY-SACHS DISEASE

Tay-Sachs disease (TSD) is a fatal, genetic (or hereditary) disorder that begins before birth and leads to gradual deterioration of the central nervous system.

The disorder actually begins early in the fetus phase of the pregnancy, even though symptoms are not apparent until the child is several months old. TSD occurs when a child lacks an important enzyme called hexosaminidase A (Hex-A). Without this enzyme, a substance called GM2 ganglioside begins to build up in abnormal amounts in the brain's nerve cells, which leads to the progressive destruction of these cells. Although a TSD baby will appear to develop normally until about six months of age, there is usually an initial loss of peripheral vision accompanied by the child exhibiting an abnormal startle response. By the age of 2, declining mental capabilities and recurrent seizures are common. A TSD infant who appears to develop normally—crawling, turning over, sitting—will eventually lose these abilities once the disease takes hold. Patients will also exhibit a loss of coordination, and have trouble swallowing and breathing. Eventually, the TSD child will become paralyzed and totally incapable of responding to his or her environment. Children with TSD usually die by the age of 5, even in the best circumstances.

Currently, there is no cure or even treatment for TSD. Attempts to provide the body with Hex-A are undergoing research, but because the disease affects brain cells, there are unique obstacles in treatment. Bone marrow transplantation (see Chapter 10) has also been researched, although many experts are wary about the chances for its success.

PARKINSON'S DISEASE

Parkinson's disease (PD) involves an impairment of the brain's basal ganglia. The neurotransmitter dopamine is important in the motor system, limbic system, and the hypothalamus. In PD, neurons located in the basal

Michael J. Fox and Parkinson's Disease

When Michael J. Fox revealed his Parkinson's disease diagnosis to the public in 1998, he already had established himself as an award-winning actor in both comedy and drama. Fox was only 30 when he began noticing symptoms of the disease in 1991. After going public with his diagnosis, he stepped away from the Hollywood spotlight in January 2000, committing himself to efforts to increase research on treatments and potential cures for Parkinson's disease. He founded the Michael J. Fox Foundation for Parkinson's Research (www.michaelj fox.org), and has said that he hopes more attention for the disease (from lawmakers and the public) will help uncover the cause of the disease and find a cure by 2010.

ganglia that produce dopamine begin to degenerate, therefore impeding mobility and other motor functions. The most common symptoms are tremors, which are involuntary shaking of the feet and hands and other limbs. Eventually, facial expressions become impossible because the facial muscles become rigid. As the disease progresses, all voluntary movements deteriorate and balance becomes impaired. In recent years, PD has been the focus of increased attention after the actor Michael J. Fox revealed his diagnosis (see "Michael J. Fox and Parkinson's Disease").

Scientists have not been able to determine the cause of PD, although some evidence suggests there is a hereditary factor. Studies have linked the onset of PD with people who took an illegal drug contaminated with the chemical MPTP (1-methyl-4-phenyl-1,2,3,6-tetrahydropyridine), in addition to people who survived a severe influenza epidemic in the early 1900s. Some scientists have also reported the identification of a certain gene that has led to PD in some people.

Increasing the levels of dopamine cannot be used to treat PD, because the neurotransmitter cannot cross the blood-brain-barrier. Instead, these transmitters are often supplemented using a drug called **L-dopa**, which can cross this barrier and be converted to dopamine by neurons in the brain. L-dopa has been shown to improve posture and motility, although it can't stop the tremors that accompany Parkinson's. In addition, L-dopa begins to lose its effectiveness within a few years. Another medication called deprenyl is reported to be successful in preventing the basal ganglia neurons from dying.

BELL'S PALSY

Bell's palsy (BP) is a form of facial paralysis that occurs when the facial nerve nucleus in the brain stem is injured, resulting in weakness to one side of the face. The injury causes neuropraxia, which means the nerves' ability to conduct impulses is slowed down. Symptoms include an inability to close the eye on the affected side of the face, difficulty smiling, and even trouble wrinkling the forehead.

Potential causes of BP include side effects from infections such as herpes simplex virus, which causes cold sores in the mouth. Current research is exploring links between BP and diabetes and HIV infection. However, in about 75 percent of BP causes, no potential cause is determined.

The two most common medicinal treatments for BP include the drug prednisone and the antiviral drug acyclovir, which is used to treat herpes. In rare cases, surgery is necessary to repair the facial nerve. The prognosis for BP patients is good, with 75 percent recovery within two to three weeks. Fifteen percent of patients experience a degree of significant recovery, although they might develop synkinesis, which means that facial nerve fibers have become somewhat misrouted during recovery. For example, when a person with synkinesis blinks, the corner of his or her mouth might slightly twitch. Some BP patients might also tear up when they eat, which is caused by a misrouting of autonomic fibers carried by the facial nerve.

STROKE

Every year, 700,000 Americans suffer a stroke, and about 190,000 of these people die. A stroke, also called a cerebrovascular accident or (CVA), is when one of the brain's blood vessels is damaged, resulting in a lack of oxygen getting to that area of the brain. The motor areas of the brain are located in the frontal lobes, where impulses are generated for voluntary motor activity and movement. The left motor area (in the left brain hemisphere) controls movement on the right side of the body, and the right motor area (on the right brain hemisphere) controls movement on the left side of the body. During a stroke, either (or both) of the motor areas might be damaged, causing muscular paralysis. If the right motor area is damaged, then the left side of the body might be paralyzed. In addition to movement, the frontal lobe is involved in reasoning, planning, emotions, and problem solving. The frontal lobe in the human brain is relatively larger than in any other organism. There are two types of blood vessel damage—thrombosis and hemorrhage.

A thrombus or blood clot often occurs as a result of atherosclerosis, which is when there are abnormal lipid deposits in the cerebral arteries. Because of these deposits, the surface of the arteries becomes rough rather than smooth. This obstructs the blood from flowing through the artery to the part of the brain it supplies. Thrombosis is when the blood clot occurs in the brain or neck, and a **stenosis** is when there is constriction of a blood vessel in the head or neck. Approximately 80 percent of strokes are caused by a thrombosis.

A hemorrhage results from an aneurysm of the cerebral artery. This causes blood to seep out into brain tissue, which puts excessive pressure on brain neurons, depriving them of oxygen and eventually destroying them. Symptoms of a hemorrhage begin quickly, and they include sudden weakness or numbness in the face, arms, or legs on one side of the body. If the stroke occurs on the left hemisphere of the brain, then the right side of the body can experience paralysis. Speech can also be damaged. When a stroke affects the vital centers in the medulla or pons, the patient can die.

For a patient who suffers from a thrombus stroke, a clot-dissolving drug can help reestablish blood flow, although it has to be given within three hours of symptom onset. Recovery from both kinds of stroke depends on where the brain is affected and the extent of the damage. It may take months of rehabilitation therapy for the patient to recover. Another treatment option involves the cerebral cortex, which has many more neurons than the average individual uses, especially in those under 50 years old. One course of rehabilitation is helping the brain to find new pathways to work by making these little-used neurons work to their full potential.

EPILEPSY

Because the brain uses electrochemical energy to operate, any disruption of this electrical process will cause abnormal function, which is what occurs when someone has epilepsy. Derived from the Greek word meaning "to possess, seize, or hold," someone who is epileptic has a brain whose neurons in the cerebral hemisphere misfire, causing abnormal electrical operation. Epileptic patients endure repeated seizures, which prevent the brain from processing incoming sensory information (such as visual and auditory). In addition, seizures prevent the brain from controlling the body's muscles. This is why someone experiencing a seizure might fall down and their muscles might twitch. Epilepsy is actually fairly common—it occurs in every 100–200 people. Napoleon Bonaparte, Julius Caesar, and Vincent van Gogh were all epileptics.

Scientists estimate that between 50 and 70 percent of epilepsy cases have an undetermined cause. For the remaining cases, brain damage resulting from a head injury, brain tumor, or stroke can cause permanent damage to the brain's electrical activity functions.

Types of Epileptic Seizures

There are numerous types of epilepsy, and each is characterized by distinct behaviors. In many cases, patients know they are about to experience a seizure when they notice an "aura," which is a kind of early warning sign such as a dizzy or nauseous feeling, that indicates a seizure is about to happen (similar to the aura that precedes some migraines).

Generalized and partial are the two most common seizures that epileptic

A French etching depicting an epileptic patient, "L'Epileptique."
© National Library of Medicine.

patients suffer. A generalized seizure occurs when there is an uncontrollable discharge of neurons on both sides of the brain. In this case, the seizure will begin in one area and then spread across the entire brain. Muscle twitches, convulsions, and loss of consciousness result. In addition, people with a generalized seizure will often not remember the episode.

There are five types of generalized seizures—tonic-clonic ("grand mal"), absence ("petit mal"), myoclonic, atonic, and status. A tonic-clonic or grand mal seizure is characterized by significant jerking (convulsing) of the body when a massive amount of neurons are discharging in both cerebral hemispheres. In contrast, an absence or petit mal seizure is nonconvulsive. Although this type of seizure only lasts between 5 and 30 seconds, the person might stare off into space and become unaware of his or her surroundings. A myoclonic seizure is convulsive, involving the motor cortex, and the atonic seizure elicits a complete loss of muscle tone, which causes the person to fall down. Finally, a status epilepticus involves frequent, long seizures accompanied by unconsciousness. This kind of seizure is extremely severe, and immediate medical attention is required.

There are two types of partial seizures—simple and complex. During a partial seizure, there is abnormal electrical activity in only a small part of the brain, although a partial seizure can sometimes spread to the whole brain, causing a generalized seizure. A simple seizure is short in duration and is without loss of consciousness. In addition, only a part of the body may convulse. A complex seizure is more significant in that the conscious state will change, the person might hear or see images, and memories may resurface.

Treatment

Although certain drugs can help control seizures, they do not cure epilepsy. Antiepileptic, or anticonvulsant, drugs change the way neurotransmitters that are responsible for seizures flow in and out of the neurons. In some cases, surgery can be used to actually treat the epilepsy. Temporal lobe surgery can be performed to remove portions of the brain tissue where the seizures begin. A corpus callosotomy might be performed to cut the corpus callosum in order to separate the brain's right and left hemispheres so the seizure does not spread. In addition, a hemispherectomy might be one option, although this is a rare option for epilepsy, because one entire cerebral hemisphere or specific lobes are removed. Research has found that children who have a hemispherectomy can function fairly well, although they might have trouble using their arm on the side of the body opposite to the surgery.

TOURETTE SYNDROME

An inherited (or genetic) disorder, Tourette syndrome (TS) is characterized by involuntary and repeated body movements, also known as **tics**. There are two types of tics—motor, which involve the face and neck muscles, and vocal, which include everything from grunting or barking to simple clearing of the throat. Many believe that most TS patients exhibit obscenity-laden outbursts, but in fact, this only occurs in 15 percent of patients. A common type of vocal tic is called echolalia, when the person repeats the words from other conversations and dialogue. The disease is named for a psychiatrist named Georges Tourette (1857–1904), who studied several patients with vocal tics, including screaming and cursing, in 1885.

TS symptoms usually occur around 7 years of age. Initial symptoms are usually facial tics, such as eye blinks, and symptoms generally decline after puberty. In fact, 20–30 percent of TS patients' symptoms completely disappear when they enter their twenties. There are three to four times more male TS patients than female ones.

Causes

Experts do not know the definitive cause of TS, although many believe it is caused in part by an abnormal gene involved with how the brain processes neurotransmitters such as dopamine, serotonin, and norepinephrine. Many researchers believe that tics are caused by receptors that are highly sensitive to dopamine in certain areas of the brain.

Because TS is inherited, research has found that if a father has TS, he has a 50 percent chance of passing it on to his child. However, the child may exhibit no symptoms at all throughout his life. Brain images have shown that TS patients display subtle abnormalities in their brain's basal ganglia.

Medication

Although most TS patients do not need medication because their symptoms are mild, many do benefit, although no single drug will cure the TS symptoms. One type of medication, **neuroleptics**, works to block dopamine receptors in the brain in order to prevent the dopamine from residing in its normal seat in the brain's nerve cells. This prevents the dopamine from transmitting signals that cause tics to the brain.

AMYOTROPHIC LATERAL SCLEROSIS

Another name for **amyotrophic lateral sclerosis (ALS)** is Lou Gehrig's disease in honor of the famous baseball player who died from the disorder in 1942. ALS patients have a working mind and memory, but their bodies will not respond to the brain's movement commands. Patients often describe the disorder as if their minds are trapped within their bodies. The cause of ALS is unknown.

ALS occurs when certain neurons located in the brain's motor cortex and the spinal cord die. These neurons control voluntary muscles and the ability to move; therefore, when these neurons die, an ALS patient may be paralyzed. Certain drugs may be used to treat the symptoms, such as muscle weakness, but there is currently no cure.

MULTIPLE SCLEROSIS

This neurological disease involves the neuron's myelin, or insulating material. In a **multiple sclerosis (MS)** patient, their myelin is damaged, and hardened areas called plaques develop along the neuron's axon. This causes multiple disruptions to the nervous system, resulting in symptoms such as difficulties in walking, visual problems, and pain. MS patients usually develop symptoms when they are 20–40 years old, and the disease is two to three times more common in women.

Although the causes of MS are unknown, research has led doctors to understand the disease's behavior, particularly its autoimmune response, which is when the body's immune system views parts of its own body as foreign invaders and attacks them. In MS, the immune system attacks the myelin surrounding the nerve cells in the brain and spinal cord. This leads to inflammation in these neurons, which then damages the myelin. Because the myelin surrounds the axon, as the myelin deteriorates, it leaves the axon exposed to damage and destruction. This subsequently damages the axons, or makes it impossible for them to transmit electrical signals. In some cases, however, the body can repair the damaged myelin, thus protecting the axon. This is referred to as remyelination.

There are four types of drugs used to treat the symptoms of MS. Three of the medications—corticosteroids, immunosuppressants, and interferons— all suppress some aspect of the immune system in order to keep the body from attacking and destroying the myelin. Corticosteroids reduce inflammation, and interferons attack antiviral agents to reduce the chance of infection. Another type of medication, glatiramer acetate, is a compound that resembles myelin. When patients are on this medication, the immune system appears to attack this compound rather than attacking the body's natural myelin.

ATTENTION DEFICIT HYPERACTIVITY DISORDER

Affecting an estimated 5 percent of school-age boys and 2 percent of school-age girls, attention deficit hyperactivity disorder (ADHD) is characterized by the inability to concentrate, along with hyperactive and impulsive behavior. Hyperactivity can include not being able to sit still, talking nonstop, and leaving one's seat when sitting is expected. Inattention includes failing to grasp certain details and rarely following directions thoroughly. Finally, impulsivity means making inappropriate comments, hitting others, and exhibiting endangering behavior, such as running out into a busy street.

The cause of ADHD has not been determined; however, research suggests that the disorder involves a deficient supply of dopamine or an inefficient functioning of dopamine receptors in the brain's anterior frontal cortex, which controls cognitive processes such as focusing. Additional research also suggests a defect in the enzyme "dopa decarboxylase," which helps make dopamine. Evidence also suggests that ADHD has genetic (or hereditary) components. In addition, brain images of ADHD boys have shown an abnormal increase of activity in the frontal lobe and nearby areas, which work to control voluntary actions. This suggests that ADHD boys actually work harder to control their impulses in comparison to non-ADHD boys.

Stimulant medications, such as Ritalin and Dexedrine, have been found to improve attention and the ability to focus in ADHD boys. These medications can be addictive in teenagers and adults, but they have not been found to be habit-forming in children. Ritalin works by reducing dopamine uptake in the brain, thus making the neurotransmitter more available in the brain for a longer period of time. Ritalin has been shown to improve symptoms such as hyperactivity and impulsivity. However, Ritalin has some unpleasant side effects, such as a decreased appetite, inability to sleep, and depression. Research is looking for ways to target the affected areas of the brain instead of increasing dopamine production in the entire brain.

AUTISM

Autism is defined as a "pervasive" development disorder, which indicates that the disorder affects many different areas and aspects of one's neurological development, including speaking. Symptoms of autism vary greatly from patient to patient, although they can be organized into three categories:

communication problems, repetitive motions, and difficulties with social interactions.

Many autistic people do not speak, make facial expressions, or make any sort of gestures, which characterizes them as incommunicative. If they do speak, they may talk in a monotone (without normal variation in their pitch), or they may speak for a long time without engaging in any sort of conversation with others. Most people who are autistic like repetitive motions, such as spinning objects. This indicates that a sense of routine is very important and that any disruption, no matter how trivial, can be extremely disconcerting. For example, if an autistic person is used to driving to the grocery store one way, he or she will be extremely upset—sometimes even inconsolable—if there is construction or anything impeding the normal route. Problems with social interactions include not making eye contact and the inability to interpret or read other people's expressions. Autistic people might also be hypersensitive to sounds such as a watch ticking. In addition, some autistic people might be extremely sensitive to touch, meaning they will avoid rough fabrics and only wear soft clothing.

Scientists have not been able to determine a cause for autism, even though it is the third most common developmental disorder in the United States, affecting at least 500,000 people. But brain images from MRI (magnetic resonance imagining) tests show that some autistic people exhibit brain abnormalities. Examples include size reduction in some parts of the cerebellum while the cerebral lobes and ventricles are larger than normal. In addition, MRI tests have shown that autistic people have smaller and more tightly packed neurons in the brain's hippocampus and amygdala.

Autism is a lifelong disorder, and there is currently no cure, although symptoms in some children decrease with age. Specialized education programs that address the behavioral and communication problems associated with autism have been shown to be successful. However, many autistic patients need institutionalized care—indeed, 50 percent never gain the ability to speak. Many autistic children go on to develop epilepsy, in addition to hyperactivity problems. In terms of medication, there has been some success with drugs that inhibit the reuptake of the neurotransmitter serotonin (or, in other words, regulate the level of serotonin in the brain). Drugs such as Fluxoetine allow the serotonin to stay in various synapses of the nervous system longer.

BIPOLAR DISORDER

Also referred to as manic depression, **bipolar disorder (BD)** is characterized by extreme changes in mood, behavior, and energy levels. BD affects nearly 1.2 percent of the U.S. population, and is thought to be hereditary. Symptoms usually appear in adolescence or early adulthood. The disorder is a cycle of behavior—from extreme highs, called **mania**, to extreme lows, called **depression**. Highs are characterized by excessive confidence, insomnia (or lack of need for sleep), distracted behavior, racing thoughts, and impulsive and excessive activities, such as spending money. Lows include a sudden loss of interest in activities normally enjoyed, changes in appetite resulting in significant weight gain or loss, changes in sleep patterns (including oversleeping), and repeated suicidal thoughts. These mood swings can also be accompanied by **psychosis** and **delusions**. Psychosis is an altered mental state when one will experience hallucinations, which means seeing and hearing things not really present. Delusions are a belief in something about oneself that is untrue, such as believing that one has the ability to fly.

Brain imaging has shown a number of changes in the brains of people with BD. These include decreases in the number and density of glial cells in the brain's prefrontal cortex, along with decreases in the number of neurons in the hippocampus. Images have also shown small abnormal areas in the brain's white matter attributed to a loss of myelin or axons, in addition to a decrease in the size of the cerebellum.

Although there is currently no cure for BD, there are medications that help control the symptoms and allow people to lead productive lives. However, medical treatment must be continuous in order to be successful. If BD is left untreated or treated sporadically, research has shown that the extreme mood episodes tend to get worse and more frequent. Many BD patients respond to Lithium, a mood-stabilizing drug that addresses the manic phase that often works in concert with other antidepressant drugs. Monoamine oxidase inhibitors, tricyclic antidepressants, and selective serotonin reuptake inhibitors are some antidepressants that regulate the amount of the neurotransmitters serotonin and norepinephrine in the brain.

CHRONIC PAIN

Pain is defined as "an unpleasant sensory and emotional experience associated with actual or potential tissue damage" according to the International Association for the Study of Pain. At its most fundamental level, pain is uncomfortable or serves as a warning against injury—such as when one has a headache after a tough day at school or work or when one's hand touches a hot stove. However, severe cases of pain can impact one's productivity, including work and personal relationships, and can adversely affect one's well being. Pain is classified into two categories—acute and chronic. **Acute pain** generally results from disease or injury that causes inflammation and related damage to the body's tissues. This type of pain is often sudden and immediately follows trauma or surgery. Acute pain is considered to be self-limiting, meaning that once the cause is identified, the pain can be treated and alleviated. **Chronic pain** is often resistant to treatment, in addition to being persistent and long-term. Because it is long term and sensory-related, chronic pain is considered a disease of the central nervous system.

When someone experiences an ache, burn, tingle, or sting, it is because receptors on the skin are triggering an electrical impulse that travels to the spinal cord, which acts as a relay center. But before it is relayed to the brain, the pain signal can be blocked, amplified, or otherwise manipulated. One specific area of the spinal cord, the dorsal horn, is vital in receiving pain signals.

The initial destination for pain signals in the brain is the thalamus, and then the **cortex**, which is considered the center of complex thoughts. In addition to communicating messages between the brain and the body, the thalamus also stores images in the brain. This is important when someone has a limb amputated. Even though the arm or leg might be gone, an image of the limb is still in the brain, and contributes to a phenomenon known as phantom pain.

Pain Basics

There is much that scientists and researchers do not know regarding pain and pain treatment or management. What is known, however, is that the body produces pain through a process involving various neurotransmitters, which stimulate neurotransmitter receptors located on the surface of cells. For each receptor, there is a corresponding neurotransmitter. These receptors act like gates or portals on the skin—they allow pain messages to pass through the skin to other cells in the CNS. One neurotransmitter found to be involved in pain is **glutamate**. Scientists have found that mice with glutamate receptors that are blocked display a diminished response to pain.

Another neurotransmitter receptor that reacts to painful stimuli is called a **nociceptor**. Nociceptors are thin nerve fibers located in the skin, muscle, and other body tissues. When these nociceptors are stimulated, they carry pain signals to the spinal cord and then to the brain. Research has shown that nociceptors usually only respond to strong stimuli, such as a slap on the face or a pinch on the arm. However, when skin tissues are inflamed or injured in some way, nociceptors are more sensitive, which causes even mild stimuli to be perceived as painful by the CNS's sensory processes.

Various disorders of the nervous system result in chronic pain. Doctors and patients often find that although the symptoms of the disorder might be addressed, chronic pain still lingers, thus inhibiting full recovery. The following are some common pain syndromes:

Arachnoiditis. This is a condition where the arachnoid membrane, which is one of the three membranes that cover the brain and spinal cord, becomes inflamed, which can lead to progressive, disabling, and even permanent pain.

Arthritis. This disorder results from inflammation of the body's soft tissues, often in muscular joints.

Back pain. A common disability for many Americans, back pain is often associated with discs in the spine, which are the spongy, soft padding located between the spine's vertebral bones. Although these discs absorb shock and thus protect the spine, they tend to degenerate (and even rupture) over time.

Neuropathic pain. This condition results from nerve injury, either in the peripheral or central nervous system. Often described as a burning sensation, neuropathic pain results from nerve damage due to injury or disease. Many cancer treatment drugs can result in neuropathic pain.

Trauma. As a result of injuries occurring at home, at the workplace, or during sports activities, many patients damage their spinal cord and suffer intense pain, in addition to temperature sensitivity. In these patients, even a light touch can be perceived by the CNS as intensely painful, which means that some abnormality is occurring in signals communicated between the brain and spinal cord. This abnormality is known as central pain syndrome. Another type of traumatic pain disorder, thalamic pain syndrome, relates to damage to the brain's thalamus, which is the CNS's center for processing bodily sensations. Patients with multiple sclerosis, Parkinson's disease, amputated limbs, and spinal cord injuries often suffer from thalamic pain syndrome.

Treatment

For headaches and other daily muscular-related aches and pains, the most common treatments are over-the-counter medications, such as aspirin, acetaminophen, and ibuprofen, which are classified as analgesics. For more severe, chronic forms of pain, "superaspirins" (COX-2 inhibitors) might be prescribed, particularly for arthritis patients. These drugs work to block two enzymes known as cyclooxygenase-1 and cyclooxygenase-2, which stimulate the production of prostaglandin hormones. Prostaglandin hormones cause tissue inflammation, fever, and pain.

Surgery might be necessary to relieve pain, particularly in the case of back or other musculoskeletal-related pain. A nerve-related surgical procedure might be necessary to block or interrupt pain messages from reaching the brain. Back surgery, or a discectomy, removes a ruptured disc. Laminectomy and spinal fusion are two other types of back surgeries. In a laminectomy, only a fragment of the disc is removed, whereas a spinal fusion involves removing the entire disc and then fusing together the two vertebrae that were separated by the disc. Other pain-related operations include a rhizotomy, in which a nerve close to the spinal cord is severed, and a dorsal root entry zone operation (or DREZ), in which surgery destroys the spinal neurons causing the pain.

FETAL ALCOHOL SYNDROME

Approximately 12,000 infants in the United States are born every year with fetal alcohol syndrome (FAS). Three times this number are believed to suffer from alcohol-related neurodevelopmental disorders (ARNDs) and alcohol-related birth defects (ARBDs). FAS is a lifelong disorder and the leading known cause of mental retardation, and it is linked to the mother's alcohol consumption during pregnancy.

As the baby develops in the womb, certain areas are covered by a cell layer that eventually develops into the bones and cartilage of the face. Alcohol exposure at such early developmental stages is believed to cause premature cell death. This is why individuals with FAS, ARND, and ARBD often display distinct patterns of facial abnormalities, growth dysfunction, and evidence of CNS dysfunction. Scientists have outlined the following facial abnormalities in FAS children: a small head, undeveloped pinna (or outer ear), short nose, missing groove above the lip, pointed and small chin, small openings for the eyes, a flat face, and thin lips. Children with FAS and related disorders often suffer neurological deficits, including slow motor skills and diminished hand-eye coordination. These children have also been found to have behavioral and learning problems, including difficulties with memory, attention, and judgment.

All of these abnormalities are considered birth defects caused by a mother's alcohol consumption during pregnancy. However, it is important to note that many FAS children do not display the physical facial characteristics and growth deficiency stated above, although they do display mental impairments. The effects of alcohol on the developing brain are not only debilitating, but they are also considered permanent, and are thought to affect the corpus callosum, which connects the right and left hemispheres of the brain. As the baby develops in the womb, research has shown that alcohol interferes with the function, growth, and survival of nerve cells in the brain.

Through research, alcohol has been defined as a teratogen, which is a substance toxic or harmful to human development. Alcohol can interfere with the natural, healthy development of a baby, depending on the amount and timing of the alcohol absorbed into his or her bloodstream. Unlike other nervous system disorders, FAS and the related syndromes are 100 percent preventable when a woman abstains from alcohol during pregnancy. When a pregnant woman drinks alcohol, it is absorbed by the baby through blood vessels in the placenta. These blood vessels also supply the developing baby with nourishment and oxygen while in the womb. But if alcohol enters a woman's body, it also enters the baby's bloodstream through the placenta. Researchers and scientists have stated that there is no known safe amount of alcohol intake during pregnancy.

CREUTZFELDT-JAKOB DISEASE

A rare, although fatal, brain disorder, Creutzfeldt-Jakob Disease or CJD causes rapid dementia and decline of mental faculties, in addition to disruption in neuromuscular functions. In individuals with CJD, neurons in the CNS often appear to look like sponges and full of holes. CJD is deadly, and the average course of the disease from symptom onset to death is four to six months. The initial symptoms of CJD include insomnia, depression, confusion, strange physical sensations, and problems with memory, coordination, balance, and sight. After this initial onset, the patient will rapidly and progressively decline mentally and physically, and begin to suffer from dementia. The patient will also exhibit jerking movements, known as myoclonus. During the final stages of CJD, the patient will lose all mental and physical capabilities, and many patients eventually fall into a coma. Death is usually the result of an infection, such as pneumonia, that the patient cannot fight given his or her unconscious state.

Scientists believe that CJD is transmitted through an agent, although this transmissible agent hasn't been defined. The current theory holds that this agent is neither a virus nor another infectious agent, but rather a pathogen known as a prion, which is an abbreviated term for a proteinaceous infectious particle. These prions are believed to alter normal protein molecules so that they ignite a chain reaction that infects neighboring healthy protein molecules and turns them into toxic molecules. There is currently no cure or effective treatment for CJD. Doctors focus on making CJD patients as comfortable as possible, such as easing their pain through medications.

Research has shown that patients generally contract CJD through one of three ways: (1) The disease can occur even without an apparent cause or be sporadic; (2) the disease can be inherited; and (3) the disease can be transmitted through an infection. When there is no known infectious source and no inherited or genetic evidence, the classification is considered sporadic CJD. Approximately 10–15 percent of CJD cases are considered inherited. Researchers believe that the patient exhibits a genetic mutation involving various prion proteins that they have inherited from a parent. If one parent carries this mutation, each of his or her offspring have a 50 percent chance of inheriting the mutation.

CJD is classified among other human and animal diseases known as transmissible spongiform encephalopathies (TSEs), and is also related to bovine spongiform encephalopathy (BSE), which is more commonly known as "mad cow disease." Scientists are currently studying the possible link between exposure to animals with mad cow disease and the onset of CJD in humans. First discovered in the United Kingdom in 1986, BSE is believed to infect animals through cattle feed that has been contaminated by TSE-in-

fected cattle. Many scientists and researchers believe that the original out-
break in the United Kingdom was probably made worse through feeding
meat and bone meal from BSE-infected cows to young calves. Mad cow disease was most rampant in the United Kingdom in 1992, with over 36,000
confirmed bovine cases.

Keeping the Nervous System Healthy

This chapter will explore some basic information about maintaining a healthy nervous system, which is vital to sustaining a healthy body in order to lead a productive life. Beginning with nutritional guidelines, including recommendations on diet and exercise, this chapter will also feature information on elements harmful to the nervous system, such as nicotine and other drugs and environmental contaminants. In addition, this chapter will discuss the effect of aging on the nervous system.

NUTRITION AND THE NERVOUS SYSTEM

The brain and other aspects of the nervous system regulate appetite, and therefore play a major role in food intake and diet. Throughout the 1990s, scientists warned that an above-average number of Americans were overweight or **obese** (see "Obesity"), putting them at greater risk for developing deadly conditions, such as heart disease and **diabetes**. Therefore, appetite and nutrition, which are regulated by the brain, are important elements in leading a healthy life.

It's important to understand and become familiar with some terms that are common and basic in nutrition, and that are the components of a healthy diet (see "Daily Dose of Nutrients"). These terms are calories, protein, fat, carbohydrates, vitamins, and minerals. **Calories**, which are available in carbohydrates, proteins, and fats, are a measurement of the amount of energy released when the body's digestive system breaks down food. It's important

Obesity

A quarter of all American adults are considered obese, increasing their risk for diseases such as heart disease, high blood pressure, some forms of cancers, and diabetes. Being overweight means having an excessive amount of body weight that includes muscle, bone, fat, and water. Obesity specifically means an excess amount of body fat. Certain athletes, such as body-builders, can be classified as overweight because they have a lot of muscle mass, but they would not be considered obese.

Most health experts believe that men with more than 25 percent body fat and women with more than 30 percent body fat are obese. The **body mass index (BMI)** has also been used to measure overweightness and obesity. The BMI is calculated using a person's height and weight. BMI equals the weight in kilograms divided by the height in meters squared (BMI = kg/m^2). The following figures (Figures 9.1 and 9.2) contain useful guidelines in determining obesity and overweightness levels for children and adults.

The breakdown of weight classes are determined by how the child's BMI (or BMI-for-age) compares with 95 percent of children of the same age and gender and at a healthy weight:

Underweight   BMI-for-age < 5th percentile

At risk of overweight   BMI-for-age is between 85th percentile and 95th percentile

Overweight   BMI-for-age ≥ 95th percentile

If a child is overweight, then at least 95 percent of children of his age and gender have a lower BMI. If a child is at risk for being overweight, then between 85 percent, but no more than 94.9 percent, have a lower BMI. If a child is underweight, only 5 percent have a lower BMI.

BMI	Weight Status
Below 18.5	Underweight
18.5 - 4.9	Normal
25.0 - 29.9	Overweight
30.0 and Above	Obese

Figure 9.1. Adult BMI.
This chart outlines weight categories in relation to body mass index for adults over 20 years old.

to limit the amount of calories consumed, because the body uses only what it needs for energy but stores the remainder as fat. Therefore, if an adult consumes more calories than he or she needs, it could lead to weight gain. If the body consumes less than it needs for energy, it will turn to stored-up fat sources for energy, which can lead to weight loss.

Proteins are nutrients that are the building blocks of the immune system, in addition to maintaining and repairing body tissue such as skin, muscles, and the body's internal organs. Proteins are composed of substances called amino acids. The body needs twenty-two amino acids to function—the body can make

Age	BMI	Percentile
2 years	19.3	95th
4 years	17.8	95th
9 years	21.0	95th
13 years	25.1	95th

Figure 9.2. Child BMI.
This chart is used to map a healthy BMI throughout childhood and the teen-age years (between 2 and 20 years of age). It's normal for a child's BMI to change throughout childhood and the teenage years.

fourteen of these, but eight come from foods such as meat, eggs, cheese, and other animal and vegetable sources. Nutritionists recommend that approximately 12 percent of an adult's daily caloric intake come from protein (see Table 9.1), but this amount can vary depending on a person's lifestyle and degree of physical activity and exercise.

Fat is made of compounds called fatty acids or lipids. These fatty acids are classified as either monounsaturated, saturated, or polyunsaturated. Saturated fats are considered the unhealthiest to eat. However, nutritionists point out that a certain amount of fat is necessary in order for the body to function properly.

Another nutritional component is **carbohydrates**, which provide fuel for all the cells in the body in the form of a sugar called glucose. Carbohydrates are classified as either simple or complex. Although simple carbohydrates

Daily Dose of Nutrients

A healthy diet means the following should make up daily calorie consumption:

- 60–70 percent carbohydrates
- 12–15 percent proteins
- 25–30 percent fats, with no more than 10 percent saturated fat

TABLE 9.1. Protein Recommendations

Sedentary adult	0.4g of protein per pound
Recreational athlete	0.5g–0.75g of protein per pound
Competitive athlete	0.6g–0.9g of protein per pound
Teenage athlete	0.9g–1.0g of protein per pound
Athlete building muscle	0.7g–0.9g of protein per pound

are found in fruit, they are also found in refined sugar, one of the main ingredients in candies and snacks, which can be unhealthy. These simple carbohydrates can provide a burst of energy, or a "sugar rush," whereas complex carbohydrates provide a more steady supply of energy as they are slowly digested by the body. Complex carbohydrates are found in beans, nuts, vegetables, and whole grains. The American Heart Association recommends a decrease in refined sugars along with an increase in the amount of complex carbohydrates to 55 percent of total daily caloric intake.

Vitamins help with the body's chemical reactions and come from food. There are thirteen vitamins that the body needs, and they are classified as either soluble or fat soluble. **Soluble** vitamins include vitamin C and all the B vitamins, and fat-soluble vitamins include vitamins A, D, E, and K. These fat-soluble vitamins are easily stored in the body, but the soluble vitamin supply needs to be replenished on a daily basis through diet. Like vitamins, minerals must also come from food. **Minerals** such as calcium, potassium, and iron are vital to the body's function and operation, and must be taken in large amounts on a daily basis. Other minerals such as zinc and copper are only needed in small amounts.

Water is also necessary to the body. In fact, without it, the body can't survive. Water accounts for most body weight (between 55 percent and 65 percent). In order for the brain to function properly, it must be adequately hydrated. Because the body can't store water, it must be a significant part of the daily diet in order for replenishment.

DIETARY GUIDELINES

The U.S. Department of Agriculture (USDA) and the U.S. Department of Health and Human Services (HHS) have developed federal Dietary Guidelines for Americans, which provides the most current information from nutritional experts across the country for people (over the age of 2) to enjoy a healthy life and reduce the chances of getting certain diseases. These include information on both diet and exercise, which will be explored separately.

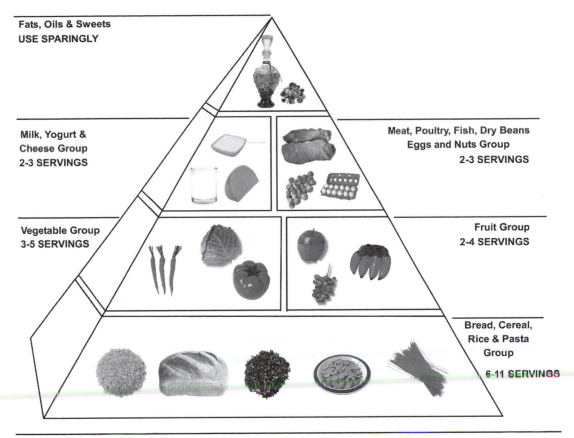

Fats, Oils & Sweets
USE SPARINGLY

Milk, Yogurt &
Cheese Group
2-3 SERVINGS

Meat, Poultry, Fish, Dry Beans
Eggs and Nuts Group
2-3 SERVINGS

Vegetable Group
3-5 SERVINGS

Fruit Group
2-4 SERVINGS

Bread, Cereal,
Rice & Pasta
Group
6-11 SERVINGS

Figure 9.3. The current Food Pyramid.
The USDA's current Food Pyramid.

There are seven official dietary guidelines:

1. Eat a variety of foods: An appropriate assortment of foods is necessary to get the appropriate amount of energy, protein, vitamins, minerals, and fiber. The USDA's Food Pyramid (Figure 9.3), which will be discussed in detail below, helps to provide this information.

2. Limit sugar intake: Sugar is full of calories and can also cause tooth decay.

3. Exercise or do some sort of physical activity on a regular basis: A physically active lifestyle reduces the chances of high blood pressure, heart disease, stroke, certain cancers, and diabetes.

4. Limit salt and sodium intake. High salt consumption is linked to high blood pressure.

5. Limit alcohol intake. Alcohol is high in calories with little nutritional value. In addition, alcohol can also lead to health problems, including addiction, and to accidents.

6. Eat a diet rich in grain products, vegetables, and fruit. This can help lower fat intake, and provides vitamins, minerals, fiber, and complex carbohydrates.

7. Eat a diet low in fat, especially low in saturated fat, and low in **cholesterol**. This will reduce risks associated with heart problems and certain types of cancer, in addition to help maintain a healthy weight.

THE FOOD GUIDE PYRAMID

The USDA's Food Pyramid was introduced in 1992 as a recommended guideline for the breakdown of daily food and serving portions (Figure 9.3). Five major food groups are shown in the lower three sections of the pyramid: the milk, yogurt, and cheese group; the meat, poultry, fish, dry beans, eggs, and nuts group; the vegetable group; the fruit group; and the bread, cereal, rice, and pasta group. It's important to eat a variety of foods to have a truly balanced diet—no one food group is more important than another. It's important to note that since 2001, this pyramid has been criticized for recommending too much refined carbohydrates, such as bread, cereal, rice, and pasta, which can cause unhealthy levels of glucose and **insulin**. An abnormally high amount of glucose and insulin in the body can lead to heart disease and other disorders of the circulatory and nervous systems (see "Food Pyramid Flaws").

The top of the pyramid shows the fats, oils, and sweets group, which need to be eaten sparingly because they are high in calories yet low in essential

Food Pyramid Flaws

When the USDA introduced the Food Pyramid guidelines in 1992, one of its main recommendations for a healthy diet was avoiding fat while eating plenty of carbohydrate-rich foods, such as bread, cereal, rice, and pasta. However, scientists and researchers have since learned that for some people, too many carbohydrates can increase glucose and insulin levels, which can lead to various heart-related diseases.

In response to these concerns, many nutritionists have proposed adopting a new Food Pyramid (Figure 9.4) that emphasizes healthy fats—polyunsaturated and monounsaturated—and whole-grain foods, such as whole-wheat or multigrain breads. This revised pyramid recommends avoiding refined carbohydrates, butter, and red meat, such as steak and hamburgers.

One of the leaders of the push to change the Food Pyramid is Dr. Walter Willett of the Harvard School of Public Health. Dr. Willett and his colleagues studied the diets of more than 100,000 men and women in the late 1990s, and found that the risk for various types of cardiovascular and nervous system–related diseases was greatly diminished through following the "new" pyramid when compared to the current guidelines.

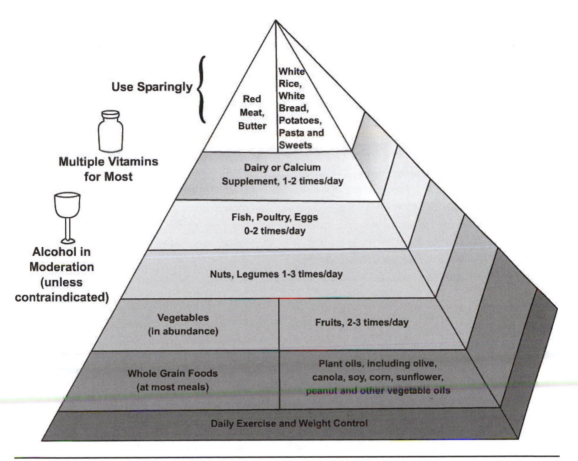

Figure 9.4. The new Food Pyramid.
The proposed new Food Pyramid. Reprinted with the permission of Simon & Schuster Adult Publishing Group from *Eat, Drink and Be Healthy: The Harvard Medical School Guide to Healthy Eating* by Walter C. Willett, M.D. © 2001 by President and Fellows of Harvard College.

nutrients. These include foods such as salad dressings and oils, cream, butter, margarine, soft drinks, candies, and desserts (see "The Lowdown on Fats and Added Sugars" for a more detailed look at fats and added sugars). Located below the fats are the milk and meat groups, which primarily come from animals and contain important nutrients such as protein, calcium, iron, and zinc. The next level—fruits and vegetables—contains foods that come from plants. Most people need to eat more from this group to increase their supply of necessary vitamins, minerals, and fiber (see "Eat 5 a Day for Better Health"). The base of the pyramid includes food from grains—breads, cereals, rice, and pasta.

Each food group in the Food Pyramid is accompanied by a range of recommended daily servings. The number of servings appropriate for each person is dependent on age, sex, size, and activity level. The National Academy

The Lowdown on Fats and Added Sugars

Although fats and sugars are concentrated at the top of the Food Pyramid, they also appear in the other food groups. For example, fruits, vegetables, and grain products are naturally low in fat. However, foods such as croissants and french fries have added fat, which diminishes their nutritional value. An average-sized baked potato has 120 calories with only a trace of fat (without butter), but an average serving of french fries has 225 calories and 11 gms of fat. In addition, foods that come from animals (the milk and meat group) are also naturally higher in fat, although there are many lowfat dairy and lean meat choices available. Added sugar can also decrease nutritional value. For instance, ice cream, chocolate milk, and cookies might be made from grain or dairy-based ingredients, but all the additional flavorings increase the calories and fat.

of Sciences in Washington, D.C., has established the following calorie guidelines for adults and teenagers (see Tables 9.2 and 9.3 for serving recommendations and diets):

- *1,600 calories*: older adults and sedentary women.
- *2,200 calories*: teenage girls, active women, and sedentary men. Women who are pregnant or breastfeeding might need more.
- *2,800 calories*: teenage boys, active men, and very active or athletic women.

KEEPING THE NERVOUS SYSTEM FIT

Another important element in addition to a healthy diet is exercise. There are various forms of physical activity—sports, scheduled exercise sessions

Eat 5 a Day for Better Health

In 1991, the National Cancer Institute (NCI) began the National 5 a Day for Better Health Program in order to encourage Americans to eat at least five to nine servings of fruits and vegetables a day. NCI scientists and researchers have determined that increasing fruit and vegetable consumption improves overall health and reduces the risk of cancer and heart problems, in addition to other diseases such as hypertension and diabetes.

TABLE 9.2. Serving Recommendations

Food Groups	Examples of Single Servings
Bread, cereal, rice, and pasta	1 slice of bread, 1 ounce of cereal, ½ cup of cooked rice or pasta
Vegetable	1 cup of raw leafy vegetable (such as spinach); ½ cup cooked, chopped, or raw vegetable; ¾ cup of vegetable juice
Fruit	1 medium apple, banana, or orange; ½ cup of cooked, chopped, or raw fruit; ¾ cup of fruit juice
Milk, yogurt, and cheese	1 cup of milk or yogurt, 1½ ounces of natural cheese, 2 ounces of processed cheese
Meat, poultry, fish, dry beans, eggs, and nuts	2–3 ounces of cooked lean meat, poultry, or fish; ½ cup of cooked dry beans or 1 egg (counts as 1 ounce of lean meat); 2 tablespoons of peanut butter or ⅓ cup of nuts (counts as 1 ounce of lean meat)

TABLE 9.3. Sample Diets

	1,600-calorie diet	2,200-calorie diet	2,800-calorie diet
Grain group servings	6	9	11
Vegetable group servings	3	4	5
Fruit group servings	2	3	4
Milk group servings	2–3	2–3	2–3
Meat group servings	5	6	7
Total fat grams	53	73	93
Total added sugars (teaspoons)	6	12	18

The Dietary Guidelines recommend limiting fat to 30 percent of calories, which calculates to the totals above. To figure this amount, multiply daily caloric intake by 0.30 to get calories from fat per day. Example: 2,200 calories × 0.30 = 660 calories from fat. Divide calories from fat per day by 9 (each gram of fat has 9 calories). Example: 660 calories from fat / 9 = 73 grams of fat.

such as aerobics or kickboxing, and even gardening or yard work can be effective ways to maintain a healthy weight and prevent certain diseases. Exercise is an important way to help control weight because it uses extra calories that would normally be stored as fat. As stated in the nutrition section, the number of calories eaten and used on a daily basis determines weight. Every piece or portion of food eaten contains calories, and every activity—even sleeping—uses calories; therefore, any additional activity to

Aerobic exercise: a young woman jogging outdoors.
© Skjold Photographs.

use up calories can lead to weight loss. When more calories are eaten than needed to perform daily activities, those calories are stored and weight is gained. However, when caloric intake decreases and the body uses those stored calories, weight can be lost. When the body uses all of the calories taken in, the weight stays the same.

Experts recommend that in order to get the most out of exercising, people should do some form of aerobic activity for 20 to 30 minutes three or more times a week. Aerobic activity is any exercise that forces one to breathe hard and use large muscle groups at an even pace. This helps to increase heart strength and use up calories. Aerobic activities include brisk walking, jogging, bicycling, swimming, aerobic dancing, racket sports, rowing, and using aerobic equipment such as a treadmill or a stationary bike (see photo).

In addition to aerobic activity, experts recommend some type of muscle-strengthening regimen (such as lifting weights) and stretching at least twice a week. It's important to check with a doctor before beginning any type of exercise routine. If exercise is a new part of a daily routine, start out with less strenuous activities such as walking or swimming at a moderate pace in order to prevent injuries.

In order to get the maximum benefits from exercise, it's important that the activity be strenuous enough to raise the heart rate to one's target zone (see Table 9.4). The middle column of the table represents the target heart zone, which is 50 percent to 75 percent of the maximum heartbeat, which is the fastest rate that the heart can beat. The heart rate can be taken by counting the number of pulse beats at the wrist or neck for 15 seconds and then multiplying it by four to get the correct beats per minute.

AGING AND THE NERVOUS SYSTEM

Some typical nervous system problems that arise with age primarily include sensory issues (hearing, vision, taste) and heart problems, which can

TABLE 9.4. Age and Target Heart Rate Zone

Age	Target Heart Rate Zone (50–75%)	Average Maximum Heart Rate (100%)
20–30 years	98–146 beats per min.	195
31–40 years	93–138 beats per min.	185
41–50 years	88–131 beats per min.	175
51–60 years	83–123 beats per min.	165
61+ years	78–116 beats per min.	155

be caused by a decreased stimulation of **vasoconstriction**, or blood flow, that is regulated by the sympathetic division (see Chapter 5 for a detailed description of the sensory systems).

Many believe that as the nervous system ages, the brain loses neurons, which leads to a decrease in mental capability and memory. This is not true. The brain does indeed lose some neurons, but the loss is only a small percentage of the total neurons. Some forgetfulness, as well as a decreased capacity for "quick thinking" or rapid problem solving, is to be expected, which is why older people generally need to drive slower and be more focused when operating a car. However, most memory capabilities remain intact in a healthy elderly person and with the proper medical attention and care, most people can minimize these undesirable aspects of aging. The main causes of dementia or mental impairment are illnesses and disorders of the nervous system such as depression, malnutrition, and heart disease. Unfortunately, many disorders such as Alzheimer's disease are currently incurable.

As the body ages, each of the senses experiences some decrease in efficacy. For example, the eye lens can become opaque and form **cataracts**, causing diminished vision. In addition, the eyes might develop presbyopia, which is when the lens' elasticity decreases and vision becomes more far-sighted.

The elderly are also at risk for **glaucoma**, an eye disorder that is caused by an increase in pressure on the eye due to an increase of aqueous humor in the canal of Schlemm. In a healthy eye, aqueous humor is produced and then reabsorbed into the canal of Schlemm, but in an eye with glaucoma, this reabsorption fails to take place, thus increasing the amount of the aqueous humor. If glaucoma is left untreated, it can lead to blindness (see Chapter 5 for more information).

Hearing can also diminish with age. After many, many years of noise, the hair cells in the organ of Corti become damaged. Usually deafness begins when high-pitched sounds cannot be heard. The ears and brain may still be able to process low-pitched sounds, although eventually this can also decrease.

The sense of taste makes eating enjoyable, and people need to eat to survive. Unfortunately, this sense tends to diminish as people age. In addition, many elderly people are on medications that tend to diminish the taste abilities. This is one possible explanation for poor nutrition among elderly and sick populations.

DRUGS AND THE NERVOUS SYSTEM

Drugs such as alcohol, nicotine, marijuana, cocaine, ecstasy, and even caffeine all have a distinct effect on the nervous system. These effects can range from minor to catastrophic and involve the nervous system's integral partners, such as the **circulatory** (heart and blood) **system**, among other parts of the body. Most of these drugs can be addictive, which means a psychological and physiological need for the substance. It is unclear whether marijuana is addictive, although many researchers are working on defining some of its addictive properties.

Alcohol

Possibly the world's oldest drug, alcohol has been produced for thousands of years. In our society, alcohol products, such as beer and wine, are sold everywhere and generate millions and millions of dollars in revenue for companies that manufacture and distribute these products. However, at the same time, alcohol abuse (alcoholism) has become a major public health issue. In fact, in the United States, transportation accidents involving automobiles, motorcycles, and bicycles are the main causes of traumatic brain injuries for people under the age of 75, and alcohol plays a role in half of all these incidents (see "Drinking and Driving" and Figures 9.5 and 9.6). Teenagers are often tempted—by their own curiosity or under pressure from their peers—to experiment with alcohol (see photo). This drug, however, can have significant adverse effects on immature bodies and minds still undergoing development, which is one reason that U.S. law prohibits alcohol consumption under the age of 21. Alcohol is considered to be a **depressant** in the central nervous system, although how it will affect a person depends on his or her age,

Peer pressure among high school students to drink alcohol. © Skjold Photographs.

Females— Approximate Blood Alcohol Concentration (BAC) Percentage

Drinks	Body Weight in Pounds									
	90	100	120	140	160	180	200	220	240	
0	.00	.00	.00	.00	.00	.00	.00	.00	.00	Only Safe Driving Limit
1	.05	.05	.04	.03	.03	.03	.02	.02	.02	Impairment Begins
2	.10	.09	.08	.07	.06	.05	.05	.04	.04	Driving Skills Significantly Affected
3	.15	.14	.11	.10	.09	.08	.07	.06	.06	
4	.20	.18	.15	.13	.11	.10	.09	.08	.08	Possible Criminal Penalties
5	.25	.23	.19	.16	.14	.13	.11	.10	.09	
6	.30	.27	.23	.19	.17	.15	.14	.12	.11	
7	.35	.32	.27	.23	.20	.18	.16	.14	.13	Legally Intoxicated
8	.40	.36	.30	.26	.23	.20	.18	.17	.15	
9	.45	.41	.34	.29	.26	.23	.20	.19	.17	Criminal Penalties
10	.51	.45	.38	.32	.28	.25	.23	.21	.19	

Subtract .01% for each 40 minutes of drinking.
One drink is 1.25 oz. of 80 proof liquor, 12 oz. of beer, or 5 oz. of table wine.

Figure 9.5. Female blood alcohol concentration.
A sample table of female blood alcohol limits used to determine intoxication. © American Academy of Pediatrics.

gender, physical health, amount of food eaten prior to consumption, and what other drugs or medications that person is taking.

Shortly after alcohol first enters the body, it moves into the stomach where a small amount enters the bloodstream. Because alcohol is a small molecule, it is soluble in water and other bodily fluid solutions. Most of the alcohol then moves to the small intestine, where it is completely absorbed in the bloodstream through the small intestine's walls. The alcohol moves through the entire body via the bloodstream and heart, and eventually

Males— Approximate Blood Alcohol Concentration (BAC) Percentage

Drinks	Body Weight in Pounds								
	100	120	140	160	180	200	220	240	
0	.00	.00	.00	.00	.00	.00	.00	.00	**Only Safe Driving Limit**
1	.04	.03	.03	.02	.02	.02	.02	.02	**Impairment Begins**
2	.08	.06	.05	.05	.04	.04	.03	.03	**Driving Skills Significantly Affected**
3	.11	.09	.08	.07	.06	.06	.05	.05	
4	.15	.12	.11	.09	.08	.08	.07	.06	**Possible Criminal Penalties**
5	.19	.16	.13	.12	.11	.09	.09	.08	
6	.23	.19	.16	.14	.13	.11	.10	.09	
7	.26	.22	.19	.16	.15	.13	.12	.11	**Legally Intoxicated**
8	.30	.25	.21	.19	.17	.15	.14	.13	
9	.34	.28	.24	.21	.19	.17	.15	.14	**Criminal Penalties**
10	.38	.31	.27	.23	.21	.19	.17	.16	

Subtract .01% for each 40 minutes of drinking.
One drink is 1.25 oz. of 80 proof liquor, 12 oz. of beer, or 5 oz. of table wine.

Figure 9.6. Male blood alcohol concentration.
A sample table of male blood alcohol limits used to determine intoxication. © American Academy of Pediatrics.

reaches the brain. At the same time, it is being broken down or oxidized by the liver into water, carbon dioxide, and energy.

See Table 9.5 for the behavioral effects of alcohol based on consumption. When alcohol is in the body, it affects many aspects of the nervous system, including the spinal cord, cerebellum, and cerebral cortex, and numerous neurotransmitters, such as norepinephrine and dopamine. Because alcohol can get into the bloodstream and blood-brain barrier quickly, it rapidly affects the

DRINKING AND DRIVING

The alcohol impairment charts for men and women (Figures 9.5 and 9.6) are a reference for approximating blood alcohol concentration (BAC). However, alcohol affects every person differently depending on weight and other variables.

brain. This alters how the body processes and produces norepinephrine and dopamine and how it transmits acetylcholine.

When alcohol consumption becomes chronic, it can lead to addiction, dependence, additional neurological problems. For the addict, withdrawal from alcohol may result in sleep problems, nausea, hallucinations, and even seizures. Other results of chronic alcohol consumption can include permanent damage to the brain's frontal lobes and overall reduction in brain size even though the ventricles will likely expand. The liver can also be permanently damaged. In addition, alcoholism can cause the body to become vitamin deficient because the body is no longer able to absorb some forms of vitamin B. This deficiency can lead to **Wernicke's encephalopathy**, which is characterized by impaired memory, confusion, and diminished coordination. Extreme deficiencies can lead to another disorder called **Korsakoff's syndrome**, which is characterized by amnesia, apathy, and disorientation. Alcohol consumption during pregnancy can also lead to fetal alcohol syndrome (FAS), discussed in Chapter 8. When a fetus is exposed to alcohol, normal brain development is disrupted, especially in the corpus callosum, which connects the right and left brain hemispheres. FAS can also result in a small basal ganglia.

TABLE 9.5. Behavioral Effects of Alcohol

Amount	Behavior
Low doses	Relaxation, tension reduction, inhibition lowered, concentration impaired, reflexes and reaction time slower, coordination impaired
Medium doses	Slurred speech, drowsiness, emotions altered
High doses	Vomiting, difficulty breathing, unconsciousness, coma

TABLE 9.6. Caffeine Contents

Drink	Size (oz.)	Caffeine (mg)
Coffee	5	60–150
Decaf coffee	5	2–5
Tea	5	40–80
Hot cocoa	5	1–8
Soft drinks	12	40–55

Caffeine

Classified as a nervous system stimulant, caffeine is probably the most popular drug in the world. It's actually hard to avoid caffeine—it's in coffee, tea, chocolate, some soft drinks, and even in some drugs and medications. Caffeine comes from the coffee bean, tea leaf, kola nut, and cacao pod (see Table 9.6 for the caffeine content of various beverages).

In moderate doses caffeine can increase alertness, although it also reduces motor coordination and alters sleep patterns, increasing the time it takes to fall asleep and reducing total sleep time. In moderate doses, caffeine can also cause headaches, nervousness, and dizziness. But in large enough doses, caffeine can actual be lethal. A fatal dose is about 10 grams, which is similar to drinking 80 to 100 cups of coffee in rapid succession.

Like alcohol, caffeine is quickly absorbed into the bloodstream through the walls of the small intestine. Once in the body, it stays there for a significant amount of time. It takes about six hours for one half the amount of caffeine in a serving of coffee to be eliminated from the body.

Caffeine belongs to the xanthine chemical group. The neurotransmitter known as adenosine is a naturally occurring form of xanthine that is found in many of the nervous system's synapses. This causes caffeine to interfere in some cases with the adenosine reception at some sites in the brain. In addition, caffeine's effects on the nervous system include increasing heart rate, constricting blood vessels, and relaxing air passages, which improves breathing and allows some muscles to contract more easily.

Some researchers and scientists have indicated that people can become physically dependent on caffeine. For example, some people feel like they can't function without a cup of coffee first thing in the morning. Many people feel that they need a midafternoon soda to keep them alert throughout the day. One way to test physical dependence is to see if someone experiences the effects of withdrawal if the caffeine source is taken away. These withdrawal symptoms could include headache, fatigue, and muscle pain, and can occur within twenty-four hours after the last dose of caffeine.

Cocaine

This drug is derived from a plant called *Erythroxylon coca,* and is a stimulant that can be smoked, inhaled (snorted), or injected. Cocaine was originally valued for its anesthetic properties. In the late 1800s, cocaine began appearing in wine and even Coca-Cola, although it was removed from the soft drink in 1906. In the United States, cocaine was made illegal in 1914, but the 1980s saw the appearance of crack cocaine, which quickly became a major drug problem.

A typical dose of inhaled cocaine is usually between 25 and 150 milligrams. Within a few seconds after it is inhaled, the user might experience euphoric feelings, excitement, feelings of strength, and a diminished appetite. These initial effects usually last about an hour, and then the users will descend or "crash" into a depressed state, leading them to use more cocaine in order to get out of this depression. Additional neurological effects during this crash can include dizziness, headaches, coordination problems, anxiety, insomnia, and hallucinations. Withdrawal from cocaine can also lead to insomnia, depression, and anxiety. Often the addicted user will feel exhausted and sleep for a long time.

Research has shown that cocaine works in the brain to block the reuptake of neurotransmitters such as dopamine, norepinephrine, and serotonin, which means these substances stay in the brain for a longer time. The peripheral nervous system is also affected, causing constricted blood vessels and an irregular heartbeat. Death by a cocaine overdose is not uncommon. Because the drug can cause a significant elevation of blood pressure, this can often lead to bleeding within the brain and constriction of the brain's blood vessels, which can lead to a stroke.

Nicotine

Tobacco, the main ingredient in cigarettes and chewing tobacco, contains the drug nicotine. Cigarette smoking causes various forms of cancer, including lip, throat, and lung cancer, in addition to respiratory problems and heart disease. American cigarettes contain about 9 milligrams of nicotine, although the smoker gets about 1 milligram of nicotine per cigarette because most of the nicotine is burned off before it reaches the body.

When tobacco smoke is inhaled, the nicotine attaches itself to small particles of tar and enters the lungs. Nicotine is absorbed quickly and reaches the brain within about eight seconds after the smoke is inhaled. Immediate effects of nicotine include an increased blood pressure and heart rate, increased rate of respiration, artery constriction, and overall stimulation of the nervous system.

Opium and Heroin

Made from the seeds of the opium poppy plant, *Papaver somniferum*, heroin is harvested in the Middle East, Southeast Asia, and parts of Central and South America. In order to make heroin, the seed pod of the opium poppy is cut, allowing a juice to flow out. **Morphine** is the main ingredient extracted from raw opium, which is converted to heroin through a chemical process.

History indicates that opium was used by the ancient Egyptians, Greeks, and Romans, and was popular to smoke in China by the 1600s. In 1680, opium was considered a medicinal therapy for pain-related health problems. In 1729, however, opium smoking was banned in China, which angered the British who controlled the trade of this drug. This began the Opium Wars of 1839–1842 between the British and Chinese involving the illegal smuggling of opium into the country. During the U.S. Civil War (1861–1865), opium was used to treated injured and sick soldiers, and was even used in cough medicine. In 1924, the Heroin Act was passed, which made the manufacture and possession of heroin illegal.

As stated earlier, heroin can be smoked, inhaled, or heated with water and injected. Because heroin is very soluble in liquids, it crosses the blood-brain barrier 100 times faster than morphine. Intravenous heroin use, which refers to injecting heroin with a syringe into a vein, produces the fastest effects—within seven or eight seconds. An overdose can lead to breathing and heart problems, or even death.

Heroin is considered a depressant because it produces an overall effect of depression in the nervous system. Short-term effects include reduced pain, or analgesia; brief euphoria; nausea; drowsiness; reduced anxiety; and ease in breathing. Long-term effects of heroin include tolerance, addiction, and withdrawal. The more heroin is used, the more is needed to produce the drug's euphoric effects. People who are addicted to heroin will often inject themselves approximately four times daily, and experience cravings within four to six hours of their last injection. Recovery is especially brutal. The first stages of recovery or withdrawal usually occur about eight to twelve hours after the last dose, and are characterized by feelings of anxiety and irritability. Several hours later, the withdrawal can escalate into sweating, fever, stomach and muscle cramps, diarrhea, and chills, which can last seven to ten days.

The exact brain mechanisms that govern addiction and tolerance are not completely understood, particularly in the case of heroin. This drug's effect on the brain involves opiate receptors, which are important for breathing, pain, and emotions. The brain produces its own opiates, like endorphins, which help a person during times of stress or pain. Opiates stimulate a kind of "pleasure system" in the brain, which involves the neurons in the midbrain that use the neurotransmitter dopamine. These midbrain neurons project stimulation to the nucleus accumbens, and then to the cerebral cortex, which creates a system of pleasure and then craving when that pleasure is diminished.

In terms of treatment, there are behavioral methods, which also include a medicinal component to alter an addict's cravings. Behavioral treatments might include positive reinforcements to stay away from the drug. But because heroin addiction is so severe, a drug called methadone has shown to be effective. Methadone blocks the effects of heroin on the brain in addition to blocking the withdrawal symptoms.

Marijuana

One of the world's most commonly used illegal drugs, marijuana comes from the plant *Cannabis sativa*. This plant contains a chemical called delta-9 tetrahydrocannabinol or THC. Marijuana can be smoked like a cigarette in a joint or used to make hash, which is composed of the marijuana plant's resin.

THC works on certain receptors in the brain that deal with memory (hippocampus), concentration (cerebral cortex), perception (sensory portions of the cerebral cortex), and movement (cerebellum). When THC reaches the brain, it disrupts normal functioning of these areas of the brain. In low or medium doses, marijuana's effects include relaxation, diminished coordination, reduced blood pressure, sleepiness, distraction, and an altered sense of time and space. In high doses, marijuana can cause hallucinations, delusions, impaired memory, and disorientation.

Marijuana can have an effect on the brain within 10 minutes after it is inhaled, and the effect can last for three or four hours. Research has shown that THC affects norepinephrine and dopamine, and may affect serotonin levels. Scientists have not determined if marijuana is addictive or what long-term mental effects THC might have on the brain.

Ecstasy

This illegal drug is also known as methylenedioxymethamphetamine or MDMA and is classified as a stimulant. Ecstasy can give people energy and increase feelings of pleasure, but headaches, chills, jaw clenching, blurred vision, and nausea are also common side effects. High doses of MDMA can also lead to dehydration and seizures.

Ecstasy appears to have three significant effects on the brain: (1) causes the release of serotonin, (2) blocks the reuptake of serotonin, and (3) decreases the amount of dopamine in the brain. Despite flooding the brain with serotonin, research shows that this can permanently damage the brain cells that release this neurotransmitter, thus affecting one's ability to remember and learn. Using brain imaging tests such as positron emission tomography (PET), studies have shown a 20–60 percent reduction in serotonin cells in ecstasy drug users. Currently, doctors do not know if this damage

Nervous System Scans

In order to get a detailed view in the nervous system, doctors will often have patients undergo screenings that work with computers to produce an image that reveals activity in the brain and the rest of the body. Below are a few types of scans commonly used.

Computerized tomography (CT): When patients undergo a CT screening, they lie on their back on a table, which moves slowly and continuously through the center of a doughnut-shaped scanner containing an x-ray beam. These x-rays scan the subject from all angles as he or she moves through the scanner, and a computer generates what will ultimately be a three-dimensional model of the lungs.

Positron emission tomography (PET): Patients who undergo this scan are injected with a radioactive material, such as oxygen, fluorine, carbon, or nitrogen, which travels through the bloodstream to the areas of the brain that use this material to function. The scan detects this radioactive material. Oxygen, for example, tends to build up in active areas of the brain. As the oxygen (or other radioactive material) breaks down, it emits both a neutron and a positron. Then when the positron hits an electron, gamma rays are produced. The PET has gamma ray detectors that then record the activity in the brain area where these gamma rays are released, producing a functional image of the brain.

Magnetic resonance imaging (MRI): This scan produces an anatomical view of the brain by detecting radio frequency signals from the brain. A patient with metallic devices, such as pacemakers, cannot undergo an MRI.

Functional magnetic resonance imaging (fMRI): This scan monitors changes in blood flow to certain areas of the brain, producing both a functional and anatomical image.

Angiography: Dye is injected into a patient's bloodstream, which enables this scan to provide images of the brain's blood vessels.

---

is permanent or if these cells will be replaced (see "Nervous System Scans" for a description of PET and other nervous system scans).

Animal studies have revealed information on how ecstasy affects animal brains, although researchers are not sure how this is significant to humans. Brain scans of animals on ecstasy have shown damage to the frontal lobe of the cerebral cortex, which is involved in thinking, and the hippocampus, which helps with memory.

Amphetamines

This category of drugs includes medicines used for sleep disorders and hyperactivity, such as Ritalin, but it can also be found in medicines used for stimulation. For example, during World War II, amphetamines were given to soldiers and pilots to combat fatigue and keep them alert.

Even though amphetamines are used to treat hyperactivity, they are actually classified as stimulants. Research shows that amphetamines increase the activity of dopamine and norepinephrine at the cells' synapses. With the use of amphetamines, dopamine levels are kept high because the drugs increase the release of the neurotransmitter from axon terminals, and also block its reuptake. In addition, these drugs inhibit both the storage of dopamine in vesicles and its enzymatic breakdown.

Amphetamines are similar to cocaine, in that tolerance increases with repeated, chronic usage. In addition, depression and fatigue are characteristic of amphetamine withdrawal, similar to cocaine withdrawal. Short-term effects include an increased heart rate and blood pressure, reduced appetite, feelings of elation and power, and reduced fatigue. Long-term effects include insomnia, paranoia, hallucinations, violent and aggressive behavior, and tremors.

Phencyclidine (PCP)

Developed in the 1950s for anesthetic purposes, PCP is classified as a dissociative drug because users appear to become disconnected from their environment when under its influence. A person might know where he or she is, but does not feel part of the surroundings. However, PCP's effects vary depending on each person. This means that based on the amount, it can act as a stimulant, depressant, or even a **hallucinogen**. Some people even become terrorized and violent when they use this drug.

Methods of taking PCP include eating, snorting, injecting, or smoking, and the drug can stay in the body for at least 10 hours. PCP affects multiple neurotransmitters, including dopamine, serotonin, opioids, and norepinephrine, mostly in a manner that blocks the brain from reuptaking those substances.

Lysergic Acid Diethylamide (LSD)

This drug was first produced in 1938 from a fungus that grows on rye and other grains by a Swiss scientist who was hoping to find a medication to stimulate circulation and respiration. Similar to PCP, LSD can have a hallucinogenic effect on a person's visual, auditory, and other sensory functions. In addition, LSD has been known to cause paranoia and a dreamlike state of consciousness.

Powerful in even small doses, LSD is odorless, colorless, and has no taste. Depending on the user's mood, the drug can produce behavioral effects including confusion, panic, happiness, distortion of time and space, chills, and muscle weakness. Although LSD's exact effect on the brain is unknown, research has shown the drug has a chemical structure similar to the neurotransmitter serotonin. In addition, studies have also given evidence that

LSD's effects are caused by the stimulation of serotonin receptors on neurons.

ENVIRONMENTAL CONTAMINANTS AND THE NERVOUS SYSTEM

Certain substances that are found naturally occurring in the environment and atmosphere can (in certain amounts) be dangerous to the body's nervous system. Some of these include lead, mercury, radiation, asbestos, and ozone.

Lead

This common metal can be found in many products such as batteries, paint, pottery glaze, and ammunition. In the past, lead has been added to gasoline to reduce engine damage, although most of the gasoline available in the United States and Europe is now available in lead-free formulas. Before 1978, lead was found in most house paint because it improved durability, but then it was banned from use in both the interior and exterior of homes. However, paint dust and chips in older homes may still contain lead. Many home water systems also contained high levels of lead because the plumbing pipes and fittings were composed of lead. The United States government passed the Safe Drinking Water Act in 1986 to restrict the amount of lead-containing materials used in distributing and producing drinking water.

Studies have shown that lead is toxic to the body and causes specific damage to the brain and peripheral nerve systems, which connect the brain and spinal cord with the rest of the body. Children are vulnerable to lead because their bodies and nervous systems are still undergoing rapid development.

Research on peripheral nerve systems poisoned by lead have shown that the myelin insulation has deteriorated and the axons are destroyed. This damage prevents messages from transmitting through the body properly, leading to problems with the sense of touch, in addition to muscle weakness. In terms of the brain, studies have shown that lead causes damage to the cerebral cortex, cerebellum, and hippocampus in addition to damaging the brain's blood vessels that leads to swelling and bleeding.

Babies who are exposed to lead in their mother's womb tend to be born premature and underweight (lead can cross the mother's blood supply to the uterus). Lead has also been shown to disrupt the development and functioning of the brain's mitochondria, which are important for energy production. Children with high levels of lead in their bodies have been shown to suffer learning disabilities and decreased growth, in addition to hyperactive behavior. Lead exposure in adults has been associated with muscle and joint pain, digestion problems, memory and concentration problems, and headaches.

Mercury

Although mercury is a useful metal that is found in products such as light bulbs and thermometers, it can also be poisonous. This contaminant comes in three main forms: elemental, organic, and inorganic. Elemental is the shiny, silver liquid used in thermometers that is absorbed by the body through vapors. Organic mercury is a form that is combined with carbon substances and is soluble in lipids, meaning that it crosses the blood-brain-barrier easily and can cause brain damage. Inorganic mercury contains non-carbon substances and is usually in a form of white powder or crystals. Organic mercury is more dangerous then inorganic mercury.

Mercury is a natural material found in the earth, and is a by-product of coal-burning power plants. It is released into the air, and then falls back to the earth. Food such as fish can be contaminated by mercury if they absorb the mineral through their gills or eat plankton or smaller fish that are contaminated. Fish such as shark, swordfish, and large tuna can contain amounts of mercury. It can also be absorbed through the skin, such as when a glass thermometer breaks. If people contain more than 50 parts-per-million (ppm) of mercury (which is measured in the hair), nerve damage could occur. Because it keeps growing, hair can be used to monitor how much mercury a person has been exposed to and when the exposure occurred.

When mercury exposure occurs, the visual cortex, cerebellum, and dorsal root ganglia are especially at risk. Mercury targets these areas and kills neurons, resulting in neurotransmitter damage in the brain and ion exchange in the cells of these neurons. Pregnant women and the children they carry are vulnerable because mercury crosses the placenta and the blood-brain-barrier, and concentrates in the fetus's brain. Because of their development stage, babies cannot excrete or get rid of the mercury. This exposure can lead to visual problems, abnormal reflexes, cognitive impairment, and disorders such as cerebral palsy.

Radiation

Naturally occurring in the earth, **radiation** is a form of energy given off by an atom's nucleus in the form of particles or rays. In terms of human exposure, the ionizing radiation can be damaging to any living tissue in the human body. Ionizing radiation comes from radioactive materials that decay spontaneously and have enough energy to create two charged ions. Even though the body will attempt to repair the damage from this type of radiation, the damage is often too widespread, leading to mistakes occurring in this natural repair process. Alpha, gamma, and beta particles, in addition to x-rays, are the most common forms of ionizing radiation.

The amount and duration of radiation that the human body is exposed to determines the severity of the tissue damage. The health effects are classified into two categories—**stochastic** and nonstochastic. Stochastic effects are related to chronic (long-term) but low-level exposure. Most scientists and researchers consider cancer to be the primary health effect from radiation. Cancer involves the uncontrolled growth of cells. This includes impact at the molecular and cellular level that disrupts the body's natural process for repairing or replacing damaged tissue, causing the cells to grow at an uncontrolled rate. The ability of ionizing radiation to break chemical bonds at an atomic and molecular level makes it a powerful **carcinogen**, or cancer-causing agent.

Additional stochastic effects include changes or **mutations** in DNA, which are the body's genetic "blueprints" that enable cell repair and replacement that produces an exact copy of the original cell. These mutations can either be teratogenic (affecting only the individual) or genetic (passed on to offspring).

Nonstochastic effects are the result of high-level exposure to radiation and become more severe as the exposure increases. Acute exposure results from short-term, high-level exposure. These effects are mainly noncancerous and appear quickly, including burns, premature aging, or even death. Symptoms of radiation illness include nausea, hair loss, skin burns, and decreased function of various organs. **Chemotherapy**—a type of treatment used to fight cancer—is a type of radiation that often causes these effects due to acute exposure.

Radiation is measured in humans in units called "rem," and many scientists estimate that the average American is exposed to a dose of about 360 millirem of radiation every year. Eighty percent of this exposure comes from natural sources such as radon gas, and rocks and soil. The remaining amount—20 percent—comes from manmade radiation sources such as medical x-rays.

Asbestos

Used in building materials such as roofing articles, electric insulation, textiles, and many other fire-resistant products, in 1971 **asbestos** became the first material to be regulated by the U.S. Dept. of Occupational Safety and Health Administration. Asbestos is the name of a group of six different minerals that are made up of long, thin fibers containing magnesium and calcium silicate. What makes asbestos distinct from other minerals is the way its crystals develop and form into these fibers. The inhalation of these fibers that make up asbestos have been linked to lung cancer and **mesothelioma**, a type of cancer that affects the membranes that line the body's abdominal cavity and surrounding organs.

The U.S. Environmental Protection Agency (EPA) states that other health risks from asbestos exposure include **pulmonary hypertension** (a disease of the lungs) and disorders of the immune system. **Asbestosis** is another lung disease caused by asbestos, and it is caused by the substance's fibers getting trapped in the lung tissue. The body tries to combat the fibers by producing an acid, but it's relatively ineffective given the chemical resistance of the fibers. This acid, however, will scar the lung tissue to the point where the lungs can no longer function. In addition, asbestos may be associated with other kinds of cancer in addition to lung cancer, including that of the esophagus, stomach, and intestines. People who work in the construction industry (especially prior to 1971) are at an especially high risk for asbestos-related illnesses. Asbestos can also be released into the water supply when pipes, roofing, or other materials containing this substance break down and disintegrate.

Ozone

Various types of air pollution, including smog, can be a health hazard because they contain dangerous levels of an odorless, colorless gas called **ozone**. This gas occurs naturally in the earth's upper atmosphere, which is 10 to 30 miles above the earth's surface. At this level, ozone forms a protective layer to shield the Earth from the sun's harmful ultraviolet rays. The "good" ozone is actually being depleted by manmade chemicals, and areas of the earth where ozone is especially thin—such as the North or South Pole—are actually said to have a "hole in the ozone." The "bad" ozone is near ground level in the earth's lower atmosphere, and is formed when pollutants from sources such as cars, power plants, industrial boilers, and chemical plants react chemically with sunlight.

Ozone has various effects on the human body, especially at high levels. One of the most common adverse health effects is irritation of the respiratory system, causing excessive coughing. Ozone at dangerous levels can also reduce lung function, disrupting normal breathing patterns. In addition to diminished lung function, ozone can also actually inflame and damage the lining of the lung, and also aggravate asthma in those who are prone to attacks. Scientists are studying the link between ozone and chronic lung diseases, such as emphysema and bronchitis.

Future Avenues of Researching the Nervous System

This chapter will focus on contemporary issues relating to research on the central nervous system (CNS), which involves discovering ways to treat disease and disorders related to the CNS. For instance, spinal cord injury research focuses on helping patients to overcome paralysis, and much of the research on Alzheimer's disease is focused on preventing and curing this debilitating disorder. Many areas of CNS research are controversial, including experimental genetics and stem cell research.

STEM CELLS

Although there are many areas of research involving the CNS, **stem cell** research is a primary aspect in the development of cell-based therapies to treat disease. This is also known as regenerative or reparative medicine, because it involves knowledge about how an organism develops from a single cell, in addition to the possibility of replacing damaged cells with healthy cells in adult organisms. Many scientists and researchers consider stem cell investigation one of the most exciting areas of research. However, embryonic stem cell research has raised ethical questions because harvesting stem cells results in the destruction of a human embryo (see "Controversies Surrounding Stem Cell Research"). Adult stem cells come from the body of a fully formed person.

Two factors distinguish stem cells from other types of cells in the human body. First, they are unspecialized, which means they do not have a specific role in the nervous system. Secondly, these cells can become specialized, or induced to perform special functions under certain experimental

Controversies Surrounding Stem Cell Research

To harvest, or produce, human stem cells in a laboratory setting, researchers must use and destroy a human embryo. Because opponents believe this is a way of destroying life, stem cell research has become embroiled in the abortion debate. Many opponents also take issue with the creation of human embryos in the laboratory for scientific research, which has also raised concerns about the ethics of this kind of research. However, supporters maintain that stem cell research could in fact hold the cure (or significant treatment advances) for diseases such as diabetes, Parkinson's, and Alzheimer's. Stem cells could be used to grow spinal cord cells, which could then be used to replace injured cells in a paraplegic. Supporters also believe that the embryos obtained from an infertility facility are eventually going to be destroyed anyway, although researchers in Virginia further fueled the debate in 2001 upon announcing that they had created human embryos in their laboratory specifically to conduct stem cell research.

Stem cell research in the United States has historically been privately funded through scientific and corporate grants. During the Clinton administration, rules for federal funding of stem cell research were written, but never implemented. When George W. Bush campaigned for office in 2000, he claimed to be against federally funding this research, although he reconsidered the issue after taking office. In 2001, President Bush eventually endorsed limited funding for the sixty existing embryo stem cell lines "where a life-and-death decision has already been made," he stated. However, all the embryos of these lines had already been destroyed.

Many supporters of stem cell research are lobbying for federal funding because they say an increase in money would speed the development of therapies, thus ensuring that the United States would be at the forefront of this pioneering research. However, many opponents claim that the use of federal funding to destroy embryos for any reason is an unacceptable use of taxpayer dollars.

conditions. For example, stem cells can be manipulated to become beating cells in the heart muscle or insulin-producing cells in the pancreas.

Scientists work with embryonic and adult stem cells from humans and from animals. In the early 1980s, researchers discovered ways to derive stem cells from mouse embryos. Then in 1998, researchers discovered how to derive stem cells from human embryos. These embryos were then used to produce cells in a laboratory setting, and were labeled as human embryonic stem cells. These early embryos were created in order to research and treat infertility through **in vitro fertilization** procedures. When they were no longer needed for this purpose, they were donated (with the donors' consent) to laboratories conducting stem cell research.

Stem Cells and Parkinson's Disease Treatments

More than 2 percent of the American population is affected by the neurogenerative disorder called Parkinson's disease (see Chapter 8 for a more detailed description of this disease). Research has show that Parkinson's disease is caused by a gradual deterioration of certain neurons that produce the neurotransmitter dopamine. This results in bodily tremors, rigidity, and hypokinesia, which means diminished mobility.

Researchers believe that Parkinson's disease could be one of the first disorders to be treated successfully with stem cell transplantation because of the involvement of the specialized dopamine-producing neurons. In addition, several laboratories have generated embryonic stem cells that differentiate into cells with some of the same functions as dopamine-producing neurons.

Research has been done using mouse embryonic stem cells that have been differentiated into these specialized neurons, then transplanted into the brains of the mice models with Parkinson's. Results showed that these differentiated neurons integrated into the mice's brains, released dopamine, and subsequently improved motor function.

As far as human stem cell therapy is concerned, scientists are attempting to conduct experiments based on the mouse model. This involves producing dopamine neurons from human stem cells in a laboratory setting and then transplanting them into the brains of humans suffering from Parkinson's disease. If this proves to be successful, neurotransplantation using dopamine neurons could prove significant for Parkinson's patients.

Stem cells are vital to all living organisms. When the embryo is between three to five days old, it is called a blastocyst. At this stage, the inner mass of a blastocyst contains a small group of about thirty cells, which produce hundreds of specialized cell types that make up an organism. Stem cells in the tissues of a developing fetus produce multiple specialized cell types that make up the heart, lung, skin, and other tissues. Various living tissues in an adult contain significant populations of stem cells in the bone marrow, muscle, and brain that serve as replacements for those cells that are lost through daily living, injury, or disease.

Many scientists and researchers believe that in the future, stem cells may be part of important treatments for various CNS-related disorders, including Parkinson's disease and spinal cord injuries, in addition to cardiovascular-related disorders such as heart disease (see "Stem Cells and Parkinson's Disease Treatments" and "Stem Cell Research on ALS Mice"). Drug therapies and toxins, in addition to understanding birth defects, might also involve stem cells. However, much remains unknown about stem cells.

Stem Cell Research on ALS Mice

In 2001, researchers at Johns Hopkins University reported that cells derived from embryonic stem cells were shown to restore limited movement in an animal model of amyotrophic lateral sclerosis (ALS), which is also known as Lou Gehrig's disease (see Chapter 8 for a more detailed description of this disorder). ALS is a nervous system disorder that gradually destroys motor neurons found in the spinal cord. This leads to increasing muscle weakness, and eventual paralysis and death.

In the Johns Hopkins experiment, the mice were exposed to Sindbis virus, a disease similar to ALS in that it also destroys the spinal cord's motor neurons, leading to paralysis of the hindquarters and weakened back limbs. Scientists judged the degree of damage through measuring the mice's movement, measuring electrical activity in the nerves functioning in the hindquarters, and visually characterizing the extent of nerve damage using a microscope.

In order to see whether stem cell transplantation could restore damaged nerves and improve mobility, the scientists used embryonic germ cells that had been isolated from human fetal tissue in 1998. Previous research had shown that when these cells are preserved and cultured, they generate copies of themselves and form into clumps called embryoid bodies. Studies had shown that these embryoid bodies could produce cells that look and function like neurons under specific laboratory conditions. With this in mind, researchers were hoping that these cells could be taken in their nonspecialized state and become specialized if placed into the area of the mice's body where the spinal cord was damaged. Therefore, the next stage of the experiment involved injecting these cultured cells from the embryoid bodies into the fluid surrounding the spinal cord of the paralyzed mice.

The Johns Hopkins scientists were pleased with the results. Three months after the injection, a significant number of the mice were able to move their hind limbs. Some were even able to walk, although clumsily. The mice that did not receive the stem cell injection remained paralyzed. Following the death of some of the injected mice, researchers were surprised to find that the injected stem cells had distributed throughout the spinal fluid and continued to develop, displaying many characteristics of motor neurons in that region of the nervous system.

Despite the positive results, the Hopkins researchers have emphasized that these results are preliminary and they are not sure if the function restoration has to do with the stem cell transplantation or if the recovery is related to the trophic factors from the injected cells that helped the remaining and undamaged nerve cells repair. But the scientific community was excited about these findings, and the study represented a major step toward applying specialized stem cells from embryonic and fetal tissues in order to restore nervous system function.

There are two key areas that researchers are focusing on, which include (1) discovering how stem cells remain unspecialized and how they renew themselves, and (2) determining how stem cells become specialized.

These two areas can lead to two treatment strategies currently at the forefront of stem cell research. The first strategy involves growing differentiated cells in a laboratory setting in order to culture them, or nudge them toward a desired cell type before implantation. Another possibility is to implant them directly into the desired area, such as the brain, and rely on signals inside the body to direct their development into the right kind of brain cell. The second research strategy involves finding specific growth factors, hormones, or other signaling molecules that will help cells to live and develop. It is hoped that these factors or molecules will stimulate the stem cells to produce specialized cells to aid in recovery from disease and injury.

Scientists are pursuing both strategies to find treatments and therapies for various disorders of the central CNS. However, a significant amount of basic research must be done before effective treatments can emerge.

SPINAL CORD INJURY RESEARCH, TREATMENT, AND REHABILITATION

The clinical management of spinal cord injuries has excelled in the past fifty years, and improved treatments are helping many people survive these traumatic injuries. However, most spinal cord injuries still severely impact the quality of life, causing lifelong disabilities such as paralysis.

The field of spinal cord injury research received much attention in 1995, when the actor Christopher Reeve was paralyzed following a horseback riding accident (see "Christopher Reeve Brings Focus on Spinal Cord Injuries"). An estimated 200,000 Americans live with the crippling trauma of a spinal cord injury on a daily basis.

In the future, scientists and researchers hope to minimize the damage from a spinal cord injury in addition to improving recovery. In recent years, new technology has led to improved imaging of damage to the spinal cord and vertebrae, in addition to diagnosing and making minor realignments of certain structural problems related to the spine immediately after the injury. Doctors are also able to greater stabilize the vertebrae following the injury in order to prevent more extensive damage. New drug therapies have also minimized related damage on the cellular level, and new rehabilitation strategies have led to improved long-term recovery, even though paralysis is still incurable.

Current research in the area of spinal cord injury is focused in four areas: (1) discovering methods to keep damaged nerves and cells at the injury site alive, (2) discovering methods to replace damaged or lost nerve tissue in order to stimulate the healthy nerves to grow and regenerate, (3) research-

Christopher Reeve Brings Focus on Spinal Cord Injuries

When Christopher Reeve was paralyzed while riding in an equestrian competition in 1995, he had established himself as a respected actor and director. Following his paralysis, many say he has put a human face on spinal cord injuries, which has led to an increase in funding for spinal cord injury research in addition to some other complex diseases rooted in the central nervous system.

In 1999, Reeve became chairman of the Christopher Reeve Paralysis Foundation (www.christopherreeve.org), which works to improve the quality of life for the disabled community. He also worked with lawmakers to help pass the 1999 Work Incentives Improvement Act, which enables people with disabilities to return to work and continue to receive disability benefits, which formerly would only be available to the unemployed. He is also involved in various other organizations that promote sporting events for athletes with disabilities, in addition to advocating for education and job opportunities for the disabled.

In addition to his charitable outreach, Reeve is also committed to bringing increased funding to the field of spinal cord research, especially from the federal government. He has testified in front of the Senate in favor of government-sponsored stem cell research. Reeve has also worked to help the passage of the New York State Spinal Cord Injury Research Bill, and is also involved in supporting similar bills in New Jersey, Kentucky, Virginia, and California. This legislation allocates up to $8.5 million annually to New York spinal cord research facilities from funds collected from violations of the state's motor vehicle laws, such as speeding or traveling without wearing a seatbelt.

ing what inhibits the growth of nerve cells in the CNS, and (4) researching rehabilitation strategies that improve recovery following an injury.

In terms of clinical management and rehabilitation, current research focuses on understanding secondary damage related to spinal cord injuries, which is the related spinal cord injuries that happen in the hours following the trauma. Following the actual trauma to the spinal cord, these delayed injuries often present doctors with brief windows of opportunity to minimize the extent of damage (and resulting disability) from the injury. One area of future research involves the immune cells present in the spinal cord following an injury.

Immune Cells

Immune cells are not abundant in the spinal cord region under normal circumstances. However, after an injury, immune cells will flood the damaged area and attempt to eliminate substances foreign to the body that are perceived as dangerous. Scientists are not certain about the role that immune cells play in helping or harming recovery potential following an injury, but studies have shown that immune reactions do lead to secondary trauma.

Part of understanding the role of immune cell function in the spinal cord involves uncovering how these cells communicate with each other. Scientists are currently working to further understand the role of **cytokines**, which are considered a kind of "messenger" molecule that controls certain elements of immune cell function and also is able to influence neuron activity. One important aspect of cytokines involves their relationship to certain cell adhesion molecules that reside on the cells' surface to control immune cell movement into the brain and spinal cord. Types of immune cells that exhibit cell adhesion molecules on their surface include the blood vessels' epithelial cells. When blood vessel and immune cells come into contact with foreign molecules, they sense damaged tissue and detect cytokines. Scientists and researchers are now working to understand how this knowledge can affect injury recovery.

In addition to these immune cells, the body releases a supply of highly reactive chemicals called free radicals or **oxidants**, which attack the body's natural defense systems and many vital cell structures. The injury or trauma that damages the spinal cord also prompts the body to release an excessive supply of neurotransmitters, which leads to excitotoxicity, which refers to secondary damage from overexcited nerve cells. Many scientists and researchers are studying ways to block or reduce this oxidative damage and excitotoxicity in order to diminish damage following a spinal cord–related injury.

Oxidative damage is when oxidants attack molecules that are essential for cell function. CNS disorders such as Lou Gehrig's and Parkinson's disease, in addition to stroke, all involve some extent of oxidative damage. Thus, oxidative damage has been the focus of intense research, as scientists search to uncover the mechanisms behind the chemicals involved in oxidation.

Necrosis

Additional inquiries into cellular damage are also a significant part of spinal cord injury research. This includes studying exactly how cells die when the spinal cord is injured or damaged. Scientists attributed cell death related to a spinal cord injury to **necrosis**, which is when cells swell at an uncontrolled rate and break open. But recent studies have shown evidence that some cells die from **apoptosis**, which is a form of "cell suicide" in which damaged cells eliminate themselves in order to prevent damaging neighboring cells. Experiments on rodents have shown that blocking apoptosis following a spinal cord injury helps to improve recovery.

Most of the damage in spinal cord injuries is attributed to harm done to the cells' axons, which are the nerve fibers that transmit signals between cells. Most neuroscientists believed for a long time that the physical element of the spinal cord injury immediately caused the axons to tear and break (see Chapter 1 for a more detailed description of a neuron's axon). However, recent studies have suggested that the trauma actually results in the disruption of the movement of molecules and cells from one end of the axon to the other, meaning that the entire axon deteriorates at a slower rate than previously believed. This delayed rate of axon deterioration might allow time for intervention.

Regeneration

Despite all that is unknown about spinal cord injuries, researchers do know that nerve cells in the spinal cord below the injured area die. Not only do the nerve cells die, but the physical damage to the spinal cord also inhibits new growth and healing, which irreparably disrupts the nerve circuits that help control movement and sensory functions. In order to understand these changes and stimulate recovery, researchers are striving to learn about pathways for function recovery through **regeneration**.

In order for regeneration to be successful, nerve cells that have been damaged by the injury must survive or be replaced (or transplanted) in order for the axons to grow again and function, which includes finding their "targets," or specialized areas where they operate. Part of this function includes constructing synapses, the specific structures that connect nerve cells. Scientists have previously transplanted adult nerve cells into the injured area of the spinal cord, but without success. However, research has shown that the injured adult spinal cord is not much different from a developing fetus's spinal cord that contains cells, synapses, and axons that are in the process of specializing and finding their correct targets. With advancements in research, doctors might one day be able to manipulate these developmental processes in order to stimulate regeneration.

Animal models of regeneration are part of these investigations. Current studies involve grafting or applying pieces from healthy peripheral nerve and fetal tissue to damaged areas of the spinal cord and neutralizing or bypassing growth-inhibiting substances in the spinal cord environment. The use of fetal tissue cells in regeneration is controversial (see "Controversies Surrounding Stem Cell Research"). However, studies involving transplanted spinal cord tissue from an aborted human embryo to the spinal cord have shown encouraging results of stability, although it's too early to draw conclusions. Because of the controversy surrounding the use of fetal cells, researchers are exploring the use of **genetically engineered** cell lines to supply nerve cells for transplantation.

Another element of regeneration involves **trophic factors** (also known as growth factors), which are natural chemicals that neurons need to feed off of in order to live and grow. These chemicals serve to nourish cells to help them grow during their developing stages. Scientists have long been frustrated with the nerve cells' inability to grow following injury. In fact, scientists have identified a gene in the human body that prevents nerve cells from growing in adults. Methods to manipulate this gene and harness the benefits of the trophic factors are other areas of regeneration that researchers are investigating. Scientists have also discovered that scar tissue can form at the site of the spinal cord injury and inhibit nerve growth.

Stem Cells and Spinal Cord Injury Research

A related area of nerve cell research involves adult stem cells recently found in areas of the brain and spinal cord that are able to divide and produce additional nerves and neurons. Scientists plan on removing stem cells from a person with a spinal cord injury and then getting the cells to divide in a lab setting. These new cells would then be injected back into the spinal cord of the person. Because the injected cells come from the patient, there is less of a chance for rejection.

But the field of stem cell therapy and spinal cord injury presents significant obstacles, and many researchers predict that although stem cells are promising in terms of treatment, complete restoration of the spinal cord (through stem cells or any other method) is probably far in the future. When someone suffers a spinal cord injury, many different types of specialized cells are destroyed, including neurons that carry messages between the brain and the rest of the body. Understanding how to grow these neurons past an injury site and then reconnect them with appropriate targets is extremely difficult and is one of the most challenging aspects of spinal cord injury research. Limited restoration, including the partial use of a limb or some restoration of bladder and bowel control, is a much more achievable goal, according to many researchers.

However, scientists have made progress in identifying the signal-carrying cells that are damaged as a result of a spinal cord injury. In many injuries, the cord is not actually cut; therefore, many of the message-carrying axons of the neurons are undamaged. Unfortunately, these axons can't transmit messages because the cells that make the axon's insulating myelin sheath—called the oligodendrocytes—have been lost due to the injury. Researchers have made initials steps toward replacing these oligodendrocytes. Some studies in mice models have found that stem cells can help remyelination when oligodendrocytes derived from mouse embryonic cells are injected into a chemically demyelinated spinal cord in a mouse. This study also showed that the mice regained a limited use of their hind limbs, although researchers are unsure whether the limited restoration of movement is actually due to remyelination or an unknown trophic effect of the treatment.

Current Medications and Therapies

In 1990, the **steroid** drug methylprednisolone was introduced. Also in 1990, clinical trials and extensive studies had shown that when given within eight hours following the injury, this drug could significantly improve recovery. Methylprednisolone works to reduce damage to the cellular membranes that contributes to neuronal death following injury. In addition, the drug appears to reduce inflammation near the injury site and suppresses the exaggerated activity of immune cells that have been shown to contribute to neural damage. The patient's chances for recovery are improved because methylprednisolone helps to protect and save some nerve fibers that would otherwise be lost or irreparably damaged. Sygen, or Gm-1 ganglioside, is another drug currently being tested for its role in decreasing cell death and thus improving recovery and regeneration of injured cells.

Another form of therapy involves **neural prostheses**, which are mechanical and electronic devices (such as a hand grasp) that connect with the CNS to supplement lost or damaged motor and sensory functions. Prosthetic devices that help patients with bladder control and leg movement are in the planning and development stages.

Clinical Management and Rehabilitation

The medical care of a patient following spinal cord injury has significantly advanced in the last fifty years. For instance, spinal cord injuries suffered during World War II were often considered fatal. Following the war, emergency and rehabilitation care advances allowed many patients to survive, but there hadn't been much research done on how to determine and influence the extent of injury.

As stated earlier, the damage to the nerve fibers as a result of spinal cord injuries can result in the loss of bladder and bowel control, sexual dysfunction, lost or diminished breathing capacity, impaired cough reflexes, and spasticity, which is an exaggerated muscle contraction. Although there is some recovery of function between a week and six months following the injury, the chances for recovery diminish after six months. Clinical management and rehabilitation work to minimize the long-term disability. Currently, clinical management of spinal cord injury focuses on three areas. First, doctors must determine the extent of the injury and relieve compression or pressure on the cord. Secondly, doctors will work to adjust any misalignments of the spine, and related structural problems. Finally, doctors will stabilize the spinal cord's vertebrae in order to prevent further injury. The immediate care of a patient following the injury is key. The emergency medical service personnel who attend the patient must be able to evaluate the injury as much as possible and immobilize the patient, because many resuscitation efforts, including any movement of the patient, can lead to further injury.

A current controversial topic in the care of spinal cord injury involves surgery to reduce pressure on the cord, which some studies show works better to stabilize the injury than just traction alone. However, some studies done in the 1970s showed that surgery actually worsened the condition. Following these findings, many doctors moved away from performing this pressure-reducing surgery, although recent advances have reduced the risk of complications. Many believe that surgery performed during that 8-hour window following injury (when the drug methylprednisolone is used) helps to allow earlier movement and physical therapy, which are key to regaining as much function as possible. Testing and scanning techniques, such as computed tomography (CT) and magnetic resonance imaging (MRI), allow doctors to view detailed damage images, including contusions and any herniated discs. These detailed views have also enabled doctors to implement the use of metal plates, screws, and other devices during surgery to greater stabilize the spine.

Once a patient is stabilized, the clinical management focuses on supportive care and rehabilitation. One of the basic elements of rehabilitation is attention to specific details that prevent further complications. For example, periodically changing the patient's position in bed or in a wheelchair prevents pressure sores and helps avoid potential breathing complications. With current rehabilitation and physical therapy techniques, almost all spinal cord injury patients can regain a partial function of some muscle function, which includes maintaining flexibility in order to avoid blood clots.

Functional Electrical Stimulation

As scientists continue to search for a cure and improve clinical management and rehabilitation strategies, research is being done to maximize physical function in spinal cord injury patients. One of these research areas involves **functional electrical stimulation (FES)**, which infuses a low-level electrical current into the neuromuscular system. This current provides the stimulation in an effort to replace the brain's nerve impulses that were damaged by the injury. Doctors use this to improve movement and function, such as helping a patient to stand. There are also therapeutic uses of FES, which include strengthening muscles that have been weakened due to the injury.

Tendon transfer can also be used in conjunction with FES, although it can also be used alone. Through the use of miniaturized computers, sensors, and electrodes, some function restoration has been achieved by replacing damaged nerve systems with silicon and copper, which are other types of neural prostheses. Through surgery, a portion of an active muscle is released and attached to a paralyzed muscle in order to stimulate motion. In August 1998, an implanted FES handgrip system was introduced to help restore hand grasp function. This system combines tendon transfer with FES to create a kind of neural prosthetic. Other FES devices currently in development include methods to allow bladder and bowel control and exercise machines to increase body strength.

FUTURE RESEARCH IN BRAIN-FOCUSED PSYCHOLOGICAL DISORDERS

In patients with psychological disorders, such as schizophrenia, bipolar disorder, and depression, researchers believe that treatment needs to be centered on those areas of the brain responsible for emotional processing (see Chapter 8 for a more detailed description of these disorders).

Brain imaging is an important element of researching treatments for these disorders. Through brain scans, scientists have discovered that brain function related to blood flow or metabolism is abnormal in certain areas of the brain—such as the prefrontal cortex, basal ganglia, and temporal lobes—during periods of both mania and depression.

Through the use of a measuring technique called **functional magnetic resonance imaging (fMRI)**, scientists have been able to measure the changes in levels of blood oxygen in different areas of the brain in both healthy people and those diagnosed with mental disorders. This technique focuses on neuron activity in the brain, because as neurons become more active, their need for oxygen increases. Because oxygen is delivered to these neurons through the blood supply, an increase in demand means an increase in blood flow to the brain. The fMRI is an important way for researchers to understand how the brains of patients with disorders such as depression and schizophrenia behave differently from the brains of healthy individuals.

In terms of bipolar disorders, an important element of treatment involves medication. Although many patients are successfully treated with medications such as **Lithium** and **Valproate**, some have troubling side effects (such as weight gain or sexual dysfunction) that may dissuade the patient from taking these drugs on a regular basis. Also, some medication regimens work well for years, but then gradually lose their effectiveness. In the case of Lithium, scientists still do not fully understand how it works or why it works for some patients and not for others. Much current and future work is focused on understanding the biochemical foundation of bipolar disorders, and targeting drug therapies for treatment. Anticonvulsant drugs that are used to treat seizures associated with epilepsy have been shown to be effective in the treatment of bipolarity. Valproate is one anticonvulsant drug that has had positive results, as is carbamazepine.

Another area of investigation relating to bipolar disorder is a nutritional approach involving omega-3 fatty acids, which are found in fish oils. Preliminary research has shown that the implementation of two main types of omega-3 fatty acids with ongoing conventional psychiatric medications has been successful in avoiding acute manic and depressive episodes, and thus maintaining the patient at a more stable emotional and psychological state for a period of over four months. Although more research is needed in this area, these results open the door to the possibility of a nutritional component in the treatment of bipolar disorders.

Another mental illness, schizophrenia, affects more than 2 million Americans and typically appears in the patient's late teens or early twenties. Research through the last decade has led to the development of several new medications (with fewer side effects), which are used in conjunction with psychosocial and behavioral therapy. However, two symptoms of the disorder, social withdrawal and loss of motivation, are still inadequately treated by medication and thus are a focus for research and investigation. Studies are continually trying to uncover how various medications interact with specific neurotransmitter systems in the nervous system in order to relieve the various symptoms of schizophrenia.

Previous research has indicated that schizophrenia is genetic, which means the predisposition to develop schizophrenia is inherited (see "Genetics and Disease"). But scientists do not understand which genes are involved in schizophrenia or how the disorder is actually transmitted from parent to child. But research into genetics is helping researchers to uncover ways to monitor the development of brain and neurotransmitter systems from infancy to adolescence and through adulthood.

Brain scans are also an important element of schizophrenia research. In various imaging studies, the presence of abnormalities in the structure and function of the brain has been detected, in addition to evidence of early biochemical changes in the brain that may precede the onset of the disease symptoms. This knowledge suggests that neural circuits are likely to be involved in the onset of these symptoms.

Another avenue of schizophrenia research that has piqued the interest of many developmental neurobiologists, who study how the brain's biological functions and processes develop, deals with research evidence that schizophrenia may occur as a result of neurons forming inadequate connections during the fetal stages of development. These impaired connections could be present at birth, but might lie dormant during puberty when there is a significant amount of nerve cell reorganization in the brain. This evidence has prompted researchers to attempt to identify prenatal factors that might indicate schizophrenia, which could include in utero infections that may adversely affect development.

Another mental disorder more common than schizophrenia is depression, which affects nearly 19 million American adults. Research advancements

Genetics and Disease

In recent years, scientists and researchers have committed a large amount of time, energy, and resources into determining the genetic factors that contribute to diseases in the CNS. Currently, a draft sequence of the human **genome** is available, which can eventually provide us with a complete picture of our genetic history. This research is key to future biomedical and pharmaceutical studies of various diseases and disorders of the CNS.

Many genetic disorders are the result of a mutation of one gene. But many diseases of the CNS, including diabetes and mental illness, have a complex genetic pattern, which means that there is no one gene that determines whether one gets the disease or not. Scientists believe that diseases are the result of more than one mutation, and a number of different genes may each contribute to one's disease susceptibility, in addition to how one might be affected by environmental factors.

In the human body, there are about 3 billion pairs of DNA that contain about 30,000 to 40,000 genes, which is much smaller than scientists initially predicted. For instance, worms and flies contain about 15,000 to 20,000 genes, roughly about half of humans. Researchers in the scientific community have devoted a significant amount of time to creating a physical map of the human genome, which means organizing the genes within the human body in addition to placing navigational landmarks. This mapping process has led to the identification of about 100 disease-related genes.

in the study and diagnosis of depression have led to improved recognition, treatment, and prevention. In the last decades, studies have shed light on the genetic, neuroscientific, and clinical aspects of depression, and much of this evidence is based on the knowledge that depression is a brain-related disorder. Brain scans and imaging techniques reveal that depression is related to a breakdown of neural circuits, which are responsible for mood regulation, sleeping, and appetite. In addition, brain images have shown an imbalance of vital neurotransmitters, which are important for nerve cells to communicate with one another. In terms of genetics, studies have shown that depression, like schizophrenia, has a genetic component, and the predisposition to develop depression is related to various genes interacting with each other and environmental factors.

Antidepressant medication is a vital area of depression research, because many patients greatly benefit from appropriate drug therapies. Current antidepressant medications influence the functioning of certain neurotransmitters in the brain, primarily serotonin and norepinephrine. These medications are known as monoamines, and some older medications include tricyclic antidepressants and monoamine oxidase inhibitors, or MAOIs. Some patients avoid these medications due to undesirable side effects, such as weight gain or sexual dysfunction. But newer medications, including drugs called **selective serotonin reuptake inhibitors (SSRIs)**, have fewer side effects than older medications.

Although antidepressant medications begin to affect brain chemistry within the first dose, they can often take several weeks in order to become effective against depression. Current research suggests that medication can be better targeted once doctors have a clearer understanding of the chemical messenger pathways within the brain's neurons, in addition to the way the genes in the brain cells behave or are expressed. For instance, in the past decade, doctors have found that many patients react better to a combination of antidepressant medications rather than just one. Many find that either the side effects are diminished or the therapeutic action against the symptoms of depression is increased. However, there is currently little information to guide doctors in determining appropriate treatment combinations. In fact, trial and error is the method most doctors, such as psychiatrists, must use in order to treat patients suffering from depression.

Medication alone cannot treat depression. **Psychotherapy** is an important element of treatment for many types of depression and other mental disorders, including bipolar disorder, which actually is a kind of depression. Research has shown that **cognitive-behavioral therapy (CBT)** and **interpersonal therapy (IPT)** can help alleviate depression. CBT involves doctors helping patients to change negative thinking patterns and self-destructive behavior, which can include substance abuse of alcohol or illegal drugs such as cocaine. With IPT, patients focus on working through personal relationships that might contribute to their depression.

One of the most controversial yet most effective treatments for acute depression is **electroconvulsive therapy (ECT)**. In fact, 80 to 90 percent of patients with severe depression experience dramatic improvements with ECT. The procedure involves electrically inducing a seizure in the brain of a patient who is under general anesthesia by applying electrodes to the scalp. The practice has come under intense scrutiny in the past thirty years by some critics in the medical community who believe it is overused and causes more harm than good. Although memory loss and other cognitive problems are common, these effects are generally short lived. In recent years, modern advances in ECT have reduced side effects. In fact, researchers have found in recent years that the effectiveness of the treatment and the severity of the side effects are often related to the dose of electricity and where the electrodes are placed on the scalp. Future challenges in the study of ECT involve how best to maintain its benefits over time. Although ECT is effective, a significant number of patients will relapse when regular treatments are discontinued.

Genetic research is also an important area of depression investigation. Scientists have focused a great deal of resources in recent years on finding a single, defective gene responsible for each mental illness. Although this identification has proved extremely difficult, scientists have begun to understand that depression and other mental disorders are the result of many different genes working together. In the next decade, however, scientists hope to pinpoint the genetic definitions of these mental disorders, which means identifying each gene related to these disorders, in addition to understanding how they are organized, or sequenced.

FUTURE OF PAIN MANAGEMENT

An important area of nervous system–related research is focused on pain management. Most disorders and diseases of the human body, whether related to the nervous system or not, involve varying degrees of pain that can range from excruciating to minor discomfort. In the last decade, scientists have focused more resources on pain management in order to improve the quality of life and expand the treatment options for patients who are dealing with pain on varying levels. Many researchers believe that a better understanding of how pain is managed in the nervous system will lead to improved medications (see Chapter 8 for more information on pain disorders).

In its most basic form, pain is uncomfortable or serves as a warning against injury—such as when one has a headache after a tough day at school or work or when one's hand touches a hot stove. But in severe cases, pain can severely impact one's productivity (inhibiting work or academic progress) in addition to adversely affecting one's well being. According to the International Association for the Study of Pain, pain is defined as "an unpleasant sensory and emotional experience associated with actual or potential tissue damage." Pain is classified into two categories—acute and chronic. Acute pain generally results from disease or injury that causes inflammation and related damage to the body's tissues. This type of pain is often sudden and immediately follows trauma or surgery. Acute pain is considered to be self-limiting, meaning that once the cause is identified, the pain can be treated and alleviated. Chronic pain is often resistant to treatment, in addition to being persistent and long term.

One of the primary goals of pain research involves developing pain medications that target the areas of the brain known as the pain "switching centers," which manage the pain-related signals throughout the nervous system. Researchers hope to uncover drug-related methods that diminish pain signals or block them altogether. Interrupting and blocking these pain signals, especially in the cases of phantom pain, is an important goal in pain medication research. One medication used in pain management is morphine, which prevents pain messages from reaching the brain. Morphine has some undesirable side effects, such as sedation, and is potentially addictive. In addition, patients can become increasingly tolerant of the drug over time, meaning that higher doses are needed for adequate pain relief. Scientists are currently working on medications that have the same pain-deadening qualities of morphine, but without the side effects.

An important element in pain medication research involves brain imaging, which allows researchers to see how the brain processes pain. Positron emission tomography (PET) and functional magnetic resonance imaging (fMRI) are two types of imaging techniques that allow researchers to see how pain activates the brain, particularly the key areas of the brain's cortex, which is the layer of tissue that covers the brain.

Another important area of pain research involves the channels, which are passages along the cells' membranes that permit ions (or electrically charged chemicals) to pass into the cells. These ion channels are vital for signal transmission throughout the nervous system's network of nerves, and researchers believe that medications can be targeted at the site of signal transmission at these ion channels. Research into trophic factors also offers hope for more effective medications. Trophic factors are the nervous system's naturally occurring chemicals that help cells survive. Studies of animal models have shown that an excess of trophic factors in nerve cells can heighten pain sensitivity. Therefore, medication to identify and minimize this excess of trophic factors could be a future avenue of pain therapy.

Three other important areas of pain research have to do with molecular genetics and **plasticity**. In regards to genetics, researchers believe that various genetic mutations can determine pain sensitivity, along with behavioral responses to pain. Some people are believed to be genetically insensitive to pain, which means that these individuals have a high pain tolerance and therefore have some form of mutation that aids in cell survival. Pain studies have been done on animal models known as "knockouts," which means these animals have been genetically engineered with mutations that cause disruptions in the processing of pain information as it leaves the spinal cord and travels to the brain. Pain management medications are often tested on these knockout animals.

Plasticity refers to the reorganization of the nervous system following an injury or a tissue-damaging disease. This reorganization causes the spinal cord to be "rewired," prompting the axons of nerve cells to make new contacts in the areas of the nervous system that remain healthy and functioning. Through the plasticity, the cells' trophic factor supply is disrupted, thus affecting how the nervous system processes pain. By using a technique called a polymerase chain reaction, scientists are able to study genes and how the changes that occur after trauma can lead to chronic pain. Scientists hope to focus potential drug therapies on preventing long-term changes that begin during plasticity and ultimately lead to chronic pain conditions.

Acronyms

ADP   adenosine diphosphate

AHA   American Heart Association

ATP   adenosine triphosphate

CDC   U.S. Centers for Disease Control and Prevention

EPA   U.S. Environmental Protection Agency

HHS   U.S. Department of Health and Human Services

NAS   National Academy of Sciences

NCI   National Cancer Institute

OSHA   U.S. Department of Occupational Safety and Health Administration

USDA   U.S. Department of Agriculture

Glossary

Abducens nerve The cranial motor nerve in the eye that controls eyeball movement with the aid of the trochlear nerve.

Acetylcholine (ACh) A neurotransmitter released in the central and peripheral nervous system, specifically at neuromuscular joints.

Action potential A change in the electrical charge of a nerve cell following the transmission of a nerve impulse.

Acute pain Type of sensory discomfort usually from disease or injury that causes inflammation and related damage to the body's tissues. This type of pain is often sudden and immediately follows trauma or surgery.

Adaptation The state of sensory acclimation in which the sensory awareness diminishes despite the continuation of the stimulus.

ADP (adenosine diphosphate) A chemical substance produced through digestion and used in cell respiration and energy production.

Adrenal glands The hormone-releasing glands located above the kidneys.

Adrenaline Also known as epinephrine, a hormone produced by the adrenal glands that helps regulate the sympathetic division of the autonomic nervous system. During times of stress or fear, the body produces additional amounts of adrenaline into the bloodstream, causing an increase in blood pressure and cardiac activity.

Aerobic cell respiration A chemical process that allows the cell to produce energy from glucose and oxygen.

Afferent nerves Fibers coming to the central nervous system from the muscles, joints, skin, or internal organs.

Afferent neuron Nerve cells that carry impulses and messages to the spinal cord and brain.

After-image An image of a visual nature that exists even after the visual stimulus has ceased.

Amino acids A chemical compound that makes up proteins.

Ampulla Located in the inner ear structures, this area is an enlarged portion that contains hair cells sensitive to movement that help to control balance and equilibrium.

Amyotrophic lateral sclerosis (ALS) Also known as Lou Gehrig's disease, this progressive nervous system disorder is characterized by a degeneration of the nerve cells that control voluntary muscle functions, resulting in muscle weakness and eventually death.

Anesthetics Chemical solutions used during surgery or other medical procedures to block ions from passing through cell membranes, thus inhibiting nerve impulse conduction.

Aneurysm A sac or bubble that can form in an artery, usually located in a weak spot.

Anoxia A type of brain injury that occurs when there is a lack of oxygen to the brain, even if there is an adequate supply of blood.

Anterior cavity The second cavity of the eye located between the front of the lens and the cornea that contains the eyeball's aqueous humor, which is the eyeball's tissue fluid formed by capillaries in the ciliary body.

Antigen These substances cause the body to produce antibodies. Examples include bacteria, toxins, and foreign blood cells.

Aorta The body's largest artery, which extends from the heart's left ventricle. The aorta has four parts: ascending aorta, aortic arch, thoracic arch, and abdominal arch.

Aortic arch One of the aorta's four parts that pumps blood through the left ventricle to over the top of the heart.

Aphasia Nerve-related disorder that disrupts speech and communication abilities.

Apoptosis A type of cell death or suicide when damaged cells eliminate themselves in order to prevent damaging neighboring cells.

Aqueous humor The eye's tissue fluid contained in the anterior cavity that hydrates the lens and cornea.

Arachnoid membrane Composed of webbed connective tissue, this brain membrane is the middle layer of the meninges.

Arachnoid villi Cerebrospinal fluid is absorbed back into the blood through these projections and into large veins within the dura mater called cranial venous sinuses.

Asbestos A group of six different minerals that are made up of long, thin fibers containing magnesium and calcium silicate. Asbestos contains crystals that form into long, thin fibers.

Asbestosis A lung disease caused by the fibers of the asbestos substance getting trapped in the lung tissue.

Atherosclerosis A brain condition in which there are abnormal lipid deposits in the cerebral arteries. Atherosclerosis can cause a stroke.

ATP (adenosine triphosphate) A chemical substance produced from aerobic cell respiration that is the muscle's direct source of energy for movement.

Auditory bones Part of the ear that receives sound wave vibrations. The bones are the malleus, incus, and stapes.

Auricle Also called the pinna, this membrane is composed of cartilage covered with skin and is located in the outer ear.

Autonomic nervous system Part of the peripheral nervous system that controls the "automatic" or involuntary movements of the body's smooth muscles (found in the walls of tubes and hollow organs), cardiac muscles, and the glands.

Axon A single nerve fiber that carries impulses away from the cell body and the dendrites.

Basal ganglia These are large masses of gray matter within the white matter located in the cerebral hemispheres, and are involved with aspects of subconscious muscle activity.

Binocular vision The process by which the brain's visual areas integrate a different picture transmitted by each eye into a single picture.

Bipolar disorder (BP) Cyclical type of depression that is characterized by extreme changes in mood, behavior, and energy levels, specifically extreme highs and lows.

Bladder Membranous sacs in the body that hold fluids and gases.

Blood-brain barrier This area of the brain is formed by capillaries between the circulating blood and blood tissue. The purpose of this barrier is to prevent harmful substances present in the blood from damaging the neurons in the brain.

Bony labyrinth The bone cavity located within the inner ear's temporal bone.

Bowel A portion of the body's intestines that connects to the anus.

Brain Besides the spinal cord, the other main organ in the central nervous system. The brain is contained within the skull and regulates the activity of the nervous system.

Brain stem This area of the brain connects the cerebrum with the spinal cord and is also the general term for the area between the thalamus and the spinal cord, which includes the medulla and pons.

Broca's area The area of the brain, located in the left cerebral hemisphere's frontal lobe, that helps to control speech. Named after the French surgeon Paul Broca (1824–1880).

Calories Units of potential energy contained in food and released upon digestion.

Canal of Schlemm The small veins at the eye's cornea and iris where the aqueous humor is absorbed.

Carbohydrates Nutritional compounds that provide cellular fuel in the form of glucose.

Carcinogen A cancer-causing agent or substance.

Carotid One of the heart's arteries that detects changes in blood pH, in addition to oxygen and carbon dioxide levels.

Carotid arteries The left and right branches of the aortic arch that transport blood through the neck and then on to the brain.

Cataracts Vision impairment or blindness caused by the opacity of the eye lens.

Catechol-O-methyl transferase An enzyme that inactivates norepinephrine and also was important in the development of drug therapies for mental illness and hypertension.

Cell body Main mass of the neuron that contains the nucleus and organelles.

Central canal The spinal cord's hollow center that contains the cerebrospinal fluid.

Central deafness One of the types of deafness that occurs when the auditory areas of the brain's temporal lobes become damaged.

Central nervous system Division of the nervous system that contains the brain and the spinal cord.

Cerebellum Located towards the back of the medulla and pons, this portion of the brain is in charge of many subconscious aspects of skeletal muscle functioning, such as coordination and muscle tone.

Cerebral angiography Method developed by Nobel Prize winner António Egas Moniz in which certain substances are introduced into the brain's blood vessels that allow specific areas of the brain to be visible on an x-ray exam.

Cerebral aqueduct The tunnel that runs through the midbrain, allowing cerebrospinal fluid to travel from the third to the fourth ventricle.

Cerebral cortex This area of the brain is the gray matter located on the surface of the cerebral hemispheres. The cerebral cortex includes the brain's motor, sensory, auditory, visual, taste, olfactory, speech, and association areas.

Cerebrospinal fluid (CSF) The fluid in the spinal cord's central canal that serves as the fluid for the central nervous system. This tissue fluid circulates in and around the brain.

Cerebrovascular accident (CVA) Another name for a stroke, this occurs when one of the brain's blood vessels is damaged, resulting in a lack of oxygen getting to that area of the brain.

Cerebrum This is the largest portion of the brain and consists of the left and right cerebral hemispheres. The cerebrum controls movement, sensation, learning, and memory.

Cervical nerves The spinal nerves in the neck.

Channels Passages along the cells' membranes that permit ions (or electrically charged chemicals) to pass into the cells.

Chemoreceptors The sensory receptors in the nervous system that detect chemical changes. In addition, there are olfactory, taste, carotid, and aortic chemoreceptors that detect changes in blood composition.

Chemotherapy An acute exposure radiation treatment for cancer.

Cholesterol A substance found in animal tissues and various foods. At high levels in the bloodstream, cholesterol can lead to certain disorders of the circulatory system.

Cholinesterase inhibitors A type of medication used by Alzheimer's disease patients that increases the level of the acetylcholine neurotransmitter.

Choroid layer The eyeball's second layer, which is made up of blood vessels. In addition, this layer prevents glare by absorbing a certain amount of light within the eyeball.

Choroid plexus This capillary network helps form the cerebrospinal fluid in the brain.

Chromosomes Cellular structures composed of proteins and DNA that carry the body's hereditary information.

Chronic pain Type of severe discomfort that is persistent, long-term, and sensory-related. Chronic pain is often resistant to treatment and is considered a disease of the central nervous system.

Ciliary body Located in the outer portion of the eye's choroid layer, this circular muscle is connected to the lens's edge by suspensory ligaments and controls the lens shape.

Ciliary muscle The main portion of the ciliary body that also helps the eye adjust to light changes.

Circadian rhythm The body's 24-hour biological cycle that regulates certain activities, such as sleep, regardless of environmental conditions, including lightness and darkness.

Circle of Willis A network of arteries that supplies the brain with blood while encircling the pituitary gland. This network is formed by the two internal carotid arteries and the two basilar arteries.

Circulatory system The heart, blood, and blood vessels in the body. The blood circulates throughout the body to deliver nutrients and remove waste.

Closed head injury A traumatic brain injury classification that refers to an injury when the head suddenly strikes an object, but the object does not break through the skull.

Coccygeal nerves The spinal nerves located in the small of the back that are also the site of attachment for some of the muscles to the pelvic floor.

Cochlea Located in the inner ear, this snail-shaped structure contains hearing receptors in the organ of Corti.

Cognition Various mental processes associated with thinking, including awareness, perception, reasoning, and judgment.

Cognitive-behavioral therapy (CBT) A type of psychotherapy that focuses on reforming fatalistic thoughts and attitudes, in addition to dangerous behavior patterns, through addressing negative and delusional thoughts.

Color blindness This inability to see color is a genetic disorder and occurs when one of the three sets of cones is dysfunctional.

Computed tomography (CT) scan A type of brain imaging scan that takes multiple x-ray images simultaneously from various angles in order to show the bone, soft tissue, and cavities of the brain.

Computerized axial tomography (CAT) Another name for a computed tomography scan (CT).

Concussion Any minor injury to the head or brain that could lead to a short loss of consciousness due to a head injury.

Conduction deafness One of the three types of deafness, it occurs when one of the ear's structures cannot transmit vibrations properly.

Cones The sensory receptors located in the eye's retina that detect colors.

Conjunctiva The eye's mucus membrane that is located on the interior of the lid that covers the white portion of the eye.

Conjunctivitis An eye infection caused by the inflammation of the conjunctiva.

Contrecoup A brain injury that results from repeated shaking back and forth of the head.

Contusion A type of skull fracture that results in bruising of the brain tissue, which is caused by the swollen brain tissue mixing with blood released from broken blood vessels.

Convergence An impulse pathway where a neuron receives impulses from the nerve endings of thousands of other neurons but transmits its message to only a few other neurons.

Cornea The initial structure of the eye that refracts light rays. The cornea is the transparent anterior portion of the sclera.

Corpus callosum A band of white matter connecting the cerebral hemispheres.

Cortex The tissue layer that covers the brain.

Cranial nerves The brain's twelve pairs of nerves located in the peripheral nervous system.

Cranial venous sinuses These large veins are located between the two cranial dura mater layers and are the location where the cerebrospinal fluid is reabsorbed.

Craniosacral division Another name for the parasympathetic division of the autonomic nervous system. In this division, all the cell bodies of preganglionic neurons are located in the brain

stem and sacral segments of the spinal cord.

Cranium The bones of the skull that house the brain.

Cutaneous The sensory components of the skin.

Cutaneous senses The skin's sensory mechanisms whose receptors are located in the dermis.

Cystic fibrosis A genetic disorder that affects the respiratory system, pancreas, and sweat glands. In this disease, the affected organs will produce an abnormal level of mucus, causing chronic respiratory infections in addition to impairment of the pancreas.

Cytokines Types of proteins that are released by the immune system's cells. These proteins, such as interleukins and lymphokines, work with other cells to generate immune responses.

Cytoplasm Cellular material located between the nucleus and cell membrane.

Deafness The inability to hear properly, which is classified into three categories: conduction deafness, nerve deafness, and central deafness.

Delusions Often a symptom of mental illness, this involves holding a false belief in spite of invalidating evidence.

Dementia The decline of intellectual capabilities such as memory, concentration, and judgment, often accompanied by emotional disturbances and personality changes as a result of a brain-related disease or disorder.

Dendrites Portion of the neuron that carries nerve impulses to the cell body.

Depolarization When the electrical charges in a nerve cell reverse due to a stimulus. The rapid infusion of sodium ions causes a negative charge outside and a positive charge inside the cell membrane.

Depressant An agent, such as a drug, that reduces physiological activity.

Depression A psychiatric-related brain disorder defined by feelings of hopelessness and extreme sadness, in addition to insomnia, loss of appetite, and thoughts of death.

Diabetes A lifelong metabolic disorder involving digestive functions and the way food is broken down to produce energy.

Divergence An impulse pathway where a neuron receives impulses from a few other neurons and relays these impulses to thousands of other neurons.

DNA A nucleic acid that contains a cell's genetic or hereditary information.

Dopamine A neurotransmitter found in the motor system, limbic system, and the hypothalamus.

Dorsal root The sensory root of a spinal nerve that attaches the nerve to the posterior part of the spinal cord.

Dorsal root ganglion An enlarged portion of the spinal nerve's dorsal root that contains the sensory neuron's cell bodies.

Dura mater This fibrous connective tissue is the outermost layer of the brain's meninges.

Ear canal The second part of the outer ear that acts as a tunnel to the middle ear.

Eardrum Also known as the tympanic membrane, this portion of the ear stretches across the end of the ear canal and produces vibrations when hit with sound waves.

Effector A muscle, gland, or other organ that responds after receiving an impulse.

Efferent nerves Fibers leaving the central nervous system carrying messages to the muscles, joints, skin, or internal organs.

Efferent neuron Nerve cells that carry impulses and messages away from the spinal cord and brain to the muscles and glands.

Electroconvulsive therapy (ECT) A treatment method for various mental disorders, particularly depression, administering electric currents through the use of electrodes on the head.

Electrode An electric conductor that can carry an electric current through a cell.

Endocrine system The body's organ system that controls hormone secretion.

Endolymph The inner ear fluid in the membranous labyrinth.

Endorphins A group of hormones that bind to the brain's opiate receptors that help control emotions and reduce pain.

Epidermal growth factor (EGF) A type of protein involved in healing wounds on the skin.

Epilepsy A type of neurological disorder characterized by recurrent seizures in addition to sudden attacks of the motor and sensory functions.

Equilibrium Balance mechanisms that are regulated by inner ear structures.

Ethology The scientific field of study that focuses on observing animals in their natural environments.

Eustachian tube The ear's air passage located between the middle ear cavity and the auditory tube, or nasopharynx.

Excitatory fiber A nerve fiber that passes impulses on to other fibers.

Excitatory synapse The passing of an impulse transmission to other synapses.

Extracellular fluid The water found outside a cell that contains plasma and other tissue fluids.

Facial nerve The mixed nerve that controls facial expressions in addition to carrying sensory impulses related to taste from the tongue to the brain.

Fat A group of soft, solid, or semisolid organic compounds composed of fatty acids or lipids.

Foramina Holes in the brain where blood vessels bring nutrients and oxygen.

Fovea The depression located in the retina behind the eye lens that contains only cones.

Frontal lobes The anterior parts of the brain containing the motor areas for speech and voluntary movement.

Functional electrical stimulation (FES) Spinal cord treatment that infuses a low-level electrical current into the neuromuscular system that provides stimulation in an effort to replace the brain's nerve impulses that were damaged by an injury.

Functional magnetic resonance imaging (fMRI) This type of brain scan produces images related to blood flow changes to certain areas of the brain.

Galvanometer An instrument that detects and measures electric currents among cells.

Ganglion neurons Groups of nerve cells located outside the CNS.

Gene A unit of hereditary information found in all living organisms. Genes consist of DNA sequences located in a specific area on a chromosome. Genes also determine the unique characteristics of each living organism.

Genetic engineering The process of altering the structure of genetic material through the production and use of recombinant DNA. Insulin and human proteins have been produced with the aid of DNA.

Genome A living organism's genetic material.

Glaucoma An eye disorder caused by an increase in pressure on the eye due to an increase of aqueous humor in the canal of Schlemm.

Glia Support cells in the brain.

Glossopharyngeal nerve The mixed nerve in the throat and salivary glands that contains sensory fibers for the throat and taste from the posterior one-third of the tongue.

Glucose A form of sugar that is a necessary component (along with oxygen) in cell respiration.

Glutamate A neurotransmitter associated with pain-related impulses.

G-protein A type of protein from the cell membrane that facilitates communication between hormone receptors and effector enzymes that regulates the metabolic and hormonal processes of a cell.

Gray matter Nerve tissue located in the central nervous system containing cell bodies of neurons.

Growth hormone (GH) A hormone that increases cell division and protein synthesis. GH is released by the anterior pituitary gland.

Growth hormone–releasing hormone (GHRH) A hormone that stimulates the anterior pituitary gland to secrete the growth hormone (GH).

Gyri Folds or ridges in the cerebral cortex.

Hallucinogen A substance or drug that alters sensory perceptions.

Hematoma A brain injury that damages a major blood vessel, causing heavy bleeding into or around the brain.

Hemispheres The lateral halves of the cerebrum.

Hemorrhage A type of stroke that results from an aneurysm of the cerebral artery. This causes blood to seep out into brain tissue, which puts excessive pressure on brain neurons, depriving them of oxygen and eventually destroying them.

Hormone A substance secreted by the endocrine gland targeted for specific organs.

Hyperopia Farsighted vision in which only distant objects are seen clearly.

Hypoglossal nerve The cranial mixed nerve that controls the movement of the tongue, swallowing in the throat, and the secretions from the throat's salivary gland.

Hypothalamus This part of the brain regulates body temperature and pituitary gland secretions. The hypothalamus is located superior to the pituitary gland and inferior to the thalamus.

Hypoxia A diminished or absent oxygen supply.

Immunoglobulin therapy Treatment therapy for some nervous system disorders when small quantities of proteins that the immune system normally uses to attack invading organisms are injected intravenously into the patient at high doses.

Imprinting The process of learning visual and auditory behavior through imitation.

Inhibitory fiber A type of nerve fiber that obstructs impulse transmission to another fiber.

Inhibitory synapse An impulse transmission obstruction due to a chemical inactivator located at the dendrite of the postsynaptic neuron.

Insulin A hormone involved in processing carbohydrates and fats, including glucose.

Intermediolateral cell column Located on the thoracic level of the spinal cord, this is an extra cell column where all presynaptic sympathetic nerve cell bodies are located.

Internal carotid arteries In conjunction with the vertebral arteries, one of two pairs of arteries that supply blood to the brain.

Interneurons Neurons located entirely within the central nervous system that combine or integrate the sensory and motor impulses.

Interpersonal therapy (IP) Often called "talk" therapy, this treatment for depression and related disorders focuses on an individual's relationships and interactions with family and friends.

Intracellular fluid The water found within a cell.

Invertebrates Living organisms that do not have a backbone.

In vitro fertilization A method of fertility treatment in which a woman's egg and a donor's sperm are combined in a test tube or laboratory dish and then inserted into the woman's body to develop.

Iris Located between the eye's cornea and lens, this colored structure has two sets of smooth muscle fibers that regulate the pupil size.

Korsakoff's syndrome Caused by an extreme vitamin deficiency related to chronic alcoholism, this disorder is characterized by amnesia, apathy, and disorientation.

Lacrimal canals These ducts are located in the outer portion of the middle of the eye and are used to transport tears to the lacrimal sac.

Lacrimal glands Tear-secreting glands of the eye.

Lacrimal sacs Located in the lacrimal bone, these openings take the eye's tears to the nasolacrimal duct, which empties tears into the nasal cavity.

Larger posterior cavity Located between the lens and the retina, this eye cavity contains the vitreous humor.

L-dopa A drug used to treat patients with Parkinson's disease. This medication can cross the blood-brain-barrier and be converted to dopamine by neurons in the brain.

Limbic system Structures in the brain relating to olfaction, autonomic functions, emotion, and behavior.

Lithium A type of antidepressant medication. Lithium has been found to be especially effective for many manic-depressive patients, because it treats both the manic highs and depressive lows characteristic of the disorder.

Long-term memory Portion of the memory function that recalls stored information.

Lumbar nerves Spinal nerves related to the six vertebrae located in the small of the back.

Lysozyme An antibacterial enzyme found in tears and saliva.

Macrophages Cells that absorb waste and other harmful material.

Macula lutea The center of the retina that contains the fovea.

Magnetic resonance imaging (MRI) This is a type of brain imaging scan that uses computer programs to trace the movements of atomic nuclei after they are exposed to radio waves within a magnetic field.

Mania A symptom of bipolar disorder that refers to intense, excessive enthusiasm and desire, but which can quickly revert to depression.

Masseter muscle The muscle that closes the jaw.

Mastication muscle The muscles in the mouth responsible for chewing.

Medulla Located above the spinal cord, this part of the brain controls vital functions such as heart rate, respiration, and blood pressure.

Membranous labyrinth Located in the inner ear, this membrane lines the bony labyrinth.

Meninges The membrane is composed of connective tissue that covers the brain and spinal cord and lines the dorsal cavity.

Mesothelioma A type of cancer that affects the membranes that line the body's abdominal cavity and surrounding organs.

Midbrain This part of the brain is located between the pons and hypothalamus and controls visual, auditory, and righting reflexes.

Minerals A variety of substances characterized by a crystalline structure. Minerals such as calcium, potassium, and iron are necessary for the body to function.

Mitochondria Located in the cell's cytoplasm, these are organelles where cell respiration takes place and energy is produced.

Mixed nerves Nerves made up of both sensory and motor fibers.

Morphine The main ingredient of heroin that is extracted from raw opium, which is then converted to heroin through a chemical process. Morphine is also a pain management medication that prevents pain messages from reaching the brain.

Motor nerves A type of afferent nerve cell leaving through the front (or anterior) of the spinal cord.

Motor neurons Also called **efferent neurons**, these carry the impulses from the central nervous system to muscles and glands.

Multiple sclerosis (MS) A nervous system disorder that is characterized by the deterioration of the myelin of the brain and spinal cord cells, resulting in muscular weakness, loss of coordination, and visual and speech impairment.

Muscle spindle Related to the stretch reflex, this receptor responds to the muscle's passive stretch and contraction. The muscle spindles are parallel with the muscle fibers.

Mutation A change in the DNA sequence of a gene or chromosome that results in a new trait not found in the parent.

Myelin sheath Substance composed of fatty material that covers most axons and dendrites in the central and peripheral nervous systems in order to electronically insulate neurons from one another.

Myoclonus Jerking movements associated with the fatal brain disorder Creutzfeldt-Jakob Disease.

Myopia Nearsighted vision, or when the eyes can see near objects but not distant ones.

Nasolacrimal duct The openings that empty the tears from the lacrimal sac into the nasal cavity.

Necrosis A type of cell death when cells swell at an uncontrolled rate and break open.

Nerve A system of neurons with blood vessels and other connective tissue.

Nerve deafness One classification of deafness, which occurs when there is damage to the eighth cranial nerve or the hearing receptors located in the cochlea.

Nerve fiber The neuron including the axon and the surrounding cells. These fibers branch out at the neuron's ending, which is known as arborization.

Nerve-growth factor (NGF) A type of protein that stimulates the growth of sensory nerves and sympathetic cells.

Nerve plexus A combination of neurons from various sections of the spinal cord that serve specific areas of the body.

Nerve tissue A type of tissue that generates and transmits electrochemical impulses.

Nerve tracts A neuron group that performs a common function in the central nervous system. This grouping can be ascending (sensory) or descending (motor).

Neural prostheses A spinal cord restoration technique that replaces damaged nerve systems with silicon and copper.

Neuroglia The non-neuron-related cells located in the central nervous system.

Neurolemma Essential to the regeneration of damaged neurons in the peripheral nervous system, this is a sheath surrounding peripheral axons and dendrites and is formed by cytoplasm and the nuclei of Schwann cells.

Neuroleptics Type of Tourette syndrome medication that works with dopamine receptors in the brain.

Neuron A nerve cell that consists of a cell body, in addition to an axon and dendrites.

Neuroscience The study of the nervous system.

Neurotransmitters Chemical substances that are emitted through nerve endings to help transmit messages. In the human body, there are about eighty different neurotransmitters.

Night blindness The inability to see clearly at night or when the light is dim. This can be caused by aging or by a deficiency in vitamin A, which is used to synthesize rhodopsin in the rods.

Nociceptor Sensory receptors in the scalp that respond to pain.

Node of Ranvier Cell region located on or between the Schwann cells.

Noradrenalin A type of neurotransmitter that transports neurons throughout the various regions in the brain and spinal cord, in addition to increasing the reaction excitability in the CNS and the sympathetic neurons in the spinal cord.

Norepinephrine A hormone that causes blood pressure to rise in stressful situations.

Nucleus The cell's largest organelle that contains chromosomes and hereditary material.

Obesity A long-term, or chronic, condition characterized by an excess of body fat that increases the patient's risks for various health problems.

Occipital lobes This is the most posterior part of the brain and contains the visual areas.

Oculomotor nerve The cranial motor nerve in the eye that controls blinking and pupil dilation.

Olfactory nerve The cranial sensory nerve that transmits sensations relating to smell from the nasal region to the brain.

Oligodendrocytes A type of neuroglia that forms the neuron's myelin sheath.

Optic chiasma The crossing location of medial fibers from each optic nerve in the eye. This structure is important for binocular vision.

Optic disc The location in the retina where the optic nerve passes through. The optic disc does not contain rods or cones.

Optic nerve The eye's cranial sensory nerve that transmits visual impulses from the eye to the brain.

Orbit The cavity that holds the eyeball.

Organelles Primary components in a cell, including the nucleus, chromosomes, cytoplasm, and mitochondria.

Organ of Corti The inner ear structure in the cochlea that contains hearing receptors.

Otoliths Located in the ear's utricle and saccule, these microscopic crystals are composed of calcium and saccule and react to gravity to help regulate equilibrium.

Oxidants A substance that oxidizes (or combines with oxygen) another substance.

Ozone An odorless, colorless gas present in various types of air pollution, including smog. Ozone is formed when pollutants from sources such as cars, power plants, industrial boilers, and chemical plants react chemically with sunlight.

Pain An unpleasant sensory and emotional experience associated with actual or potential tissue damage.

Papillae The projections on the tongue that contain taste buds.

Paranoia A psychiatric disorder characterized by delusions.

Paraplegia The loss of sensation and function, often due to a spinal cord injury, to the leg and lower parts of the body.

Parasympathetic division The division of the autonomic nervous system that dominates and controls the body during nonstressful situations.

Parietal lobes The portion of the brain's cerebrum that is posterior to the front lobes and contains the sensory areas for cutaneous and conscious muscle sense.

Penetrating head injury A traumatic brain injury classification that refers to an injury when an object breaks through the skull, piercing the brain tissue.

Peptides A chemical that helps to join amino acids in a protein molecule.

Perilymph The inner ear fluid located in the bony labyrinth.

Peripheral nervous system Division of the nervous system that consists of the spinal and cranial nerves.

Peristaltic Muscular contractions that occur in a wavelike motion that force contents onward through a canal or tubular structure.

Phantom pain Following an amputation, this is the pain that seems to come from the missing limb.

Phrenology The field of study based on the idea that certain human behaviors and characteristics depend on the size, shape, and patterns of bumps on the skull.

Pia mater The meninges' innermost layer, made of thin connective tissue located on the surface of the brain and spinal cord.

Plasmapheresis Treatment therapy for some nervous system disorders, in which a portion of the patient's blood is removed and then the liquid component is cleansed. These clean blood cells are then injected into the body.

Plasticity The reorganization of the nervous system following an injury or a tissue-damaging disease.

Polarization A chemically charged state when the neuron's membrane has a positive charge outside and a negative charge inside.

Pons The parts of the brain that are anterior and superior to the medulla. The pons regulate respiration.

Positron emission tomography (PET) scan A type of brain imaging technique that shows the brain in action. In order to obtain this image, a radioactive substance (such as glucose) is injected into the brain and then followed as it moves throughout the brain.

Postganglionic neuron A neuron located in the autonomic nervous system that extends from a ganglion to the visceral effector.

Postsynaptic Any impulse event following transmission at the synapse.

Prefrontal lobotomy A kind of surgery developed in the 1940s that was thought to cure brain disorders such as paranoia by severing the nerve fibers present in the brain's frontal lobes (which are responsible for psychological responses) from the thalamus (a relay center for the brain's sensory impulses).

Preganglionic neuron A neuron located in the autonomic nervous system that extends from the central nervous system to a ganglion and then synapses with a postganglionic neuron.

Presbyopia Often due to aging, this farsighted vision condition is related to the lens's loss of elasticity.

Pressoreceptors The sensory receptors that detect changes in blood pressure in the carotid and aortic sinuses.

Pressure sores Areas on the skin where the tissue breaks down due to long periods of sitting or laying following a spinal injury.

Presynaptic Any impulse event related to before transmission at the synapse.

Prions A protein thought to be an infectious agent partly responsible for various degenerative disorders of the nervous system.

Projection A sensory occurrence when the sensation is felt in the receptor area.

Proteins A group of molecules containing carbon, hydrogen, oxygen, nitrogen, and usually sulfur that are composed of a chain of amino acids. The basic component of all living cells, proteins are necessary in order for enzymes and hormones to function. They are also vital for the growth and repair of cell tissue.

Psychosis A type of altered mental state characterized by hallucinations.

Psychotherapy A treatment method for mental disorders that focuses on communicating problems with a professional psychiatrist. The goal of psychotherapy is to gain an insight into mental and emotional processes, in addition to changing behavior and social patterns to enhance happiness and productivity.

Pulmonary hypertension A severe lung disease that has been linked to asbestos.

Pupil The opening in the center of the iris where light rays pass.

Quadriplegia The loss of sensation and function in the legs and arms, usually due to a spinal cord injury.

Radiation A form of energy given off by an atom's nucleus in the form of particles or rays. Ionizing radiation can be damaging to any living tissue in the human body.

Receptors A cell or nerve that responds to a specific change or stimulus.

Reflex An automatic or involuntary response to a stimulus.

Refraction The process by which the eye bends light in order to produce an image.

Regeneration The renewal of cells, tissues, nerves, and other parts of the body through the regrowth of epithelial cells.

Repolarization A chemically charged state following a neuron's depolarization, when the membrane has a positive charge outside and a negative charge inside due to the outflow of potassium ions.

Retina The portion of the eye that contains the rods, cones, and photoreceptors.

Rhodopsin Located in the rods of the retina, this chemical is broken down by light rays, which causes a chemical change that initiates a nerve impulse.

Rods The sensory receptors located in the retina that detects light rays.

Saccule The sac in the inner ear that contains equilibrium receptors.

Sacral nerves The spinal nerves located at the base of the spine.

Sarcolemma The cell membrane of a muscle fiber.

Schizophrenia A group of psychiatric-related disorders that involve isolation and a withdrawal from reality, irrational cognitive processes, hallucinations, in addition to related emotional and behavioral disturbances. These disorders are believed to be associated with dopamine imbalances in the brain, in addition to defects in the brain's frontal lobe.

Schwann cells Located in the peripheral nervous system, these cells form the myelin sheath and neurolemma of the peripheral axons and dendrites.

Sclera Composed of fibrous connective tissue, this is the outer layer of the eyeball.

Seizures A sudden attack or convulsion related to epilepsy or other disorder.

Selective serotonin reuptake inhibitors (SSRIs) Antidepression medications that control neurotransitter levels in the brain.

Semicircular canals Located in the inner ear, these three canals contain motion receptors.

Sensory nerves A type of afferent nerve coming in at the back of the spinal cord; also called posterior nerves.

Sensory neurons Also known as afferent neurons, they carry impulses and messages to the spinal cord and brain.

Sensory tract The pathways located in the brain or spinal cord's white matter, through which impulses travel to reach a specific area of the brain.

Serotonin A neurotransmitter present throughout the central nervous system.

Sherrington's Law Named for 1932 Nobel Prize winner Sir Charles Scott Sherrington (1857–1952), this law states that when one set of muscles is stimulated, the opposite muscles, or those opposing the action, are inhibited.

Short-term memory Portion of the memory function that retains small amounts of information only for a short time.

Skull The brain's bony encasement.

Skull fracture An injury that occurs when the skull cracks or breaks.

Soluble The ability to be dissolved in a liquid such as water.

Somatic neuron A type of sensory neuron located in the skeletal muscle and joints.

Spasticity Following a spinal injury, these spasms and exaggerated reflexes often occur in areas no longer controlled by the brain.

Specific nerve energy theory First proposed by Johannes Müller (1801–1856), this theory stated that different nerves were coded to carry out a specific function, which depended on where they originated in the brain.

Spinal accessory nerve The cranial motor nerve that controls the throat muscles and the two major neck muscles.

Spinal cord Located in the vertebral canal, this is an organ that transmits impulses to and from the brain.

Spinal nerves The spine's thirty-one pairs of nerves located in the peripheral nervous system.

Spinal reflex An automatic or involuntary reflex related to the spinal cord and in which the brain is not directly involved.

Stem cell An unspecialized cell that can be altered to form a specific (or specialized) function.

Stenosis A constriction of a blood vessel in the head or neck.

Steroid A carbon-containing substance that composes important elements of hormones, such as testosterone and estrogen. Steroids are also important in the production of cortisone, which is excreted from the adrenal gland.

Stimulus Any sort of change in a living organism that causes a response or affects a sensory receptor.

Stochastic Effects related to chronic, low-level exposure of radiation.

Stretch receptors A sensory receptor located in the muscle that detects when the muscle is stretching.

Stretch reflex A reflex from the spinal cord in which a muscle will respond to a stretch by contracting.

Stroke This is also called a cerebrovascular accident and is when one of the brain's blood vessels is damaged, resulting in a lack of oxygen getting to that area of the brain. There are two types of blood vessel damage—thrombosis and hemorrage.

Sulci Grooves between the gyri of the cerebellum.

Sumatriptan Family of migraine medications used to counteract the dilation and inflammation of the brain's blood vessels associated with severe headaches.

Sympathetic division The division of the autonomic nervous system that dominates and controls the body during stressful situations.

Synapse The junction between two neurons where the axon passes on information to the dendrite. This area is often called a relay because it is here where the information is relayed to the next neuron.

Synaptic gap or **cleft** The actual area (which is approximately 10–50 nanometers in width) between the axon and dendrite where the neurons communicate with each other.

Temporal lobes The lateral parts of the cerebrum that contain the auditory, olfactory, and taste areas.

Thalamus The portion of the brain located superior to the hypothalamus that controls the elements of subconscious sensation.

Thoracic nerves The twelve spinal nerves related to the rib area of the spinal cord.

Threshold level This value in a nerve fiber depends on the composition of the cellular fluid and the number of impulses recently received and conducted. When this level is reached in the nerve fiber's axon, a reaction results.

Thrombosis A blood clot occurring in the brain or neck that can cause a stroke.

Thrombus A type of stroke that is caused by a blood clot and is a result of abnormal lipid deposits in the cerebral arteries.

Thyrotropin-releasing hormone (TRH) A hormone that stimulates the thyroid gland, which is involved with maintaining normal growth and metabolism rates.

Tics Often occurring in the face, these are involuntary muscular contractions or spasms.

Tourette syndrome A neurological disorder characterized by verbal outbursts and utterances, in addition to multiple facial and body tics.

Traumatic brain injury (TBI) A head injury that impairs the brain.

Trigeminal nerve The cranial mixed nerve from the face and mouth that controls chewing and sensations by carrying motor fibers from the face and mouth to the mastication muscles.

Trochlear nerve The cranial motor nerve that controls eye movement with the aid of the abducens nerve.

Trophic factors Natural chemical substances in the body's nervous system that help cells to function and survive.

Utricle Located in the inner ear's vestibule, this membrane contains equilibrium receptors.

Vagus nerve The cranial mixed nerve that controls the movements and sensations in the organs located in the rib and abdominal region, such as the glands, digestion, heart rate, and the voice box.

Valproate A type of anticonvulsant medication found to be effective in the treatment of bipolar disorder.

Vascular headaches Also called migraines, these are headaches that involve abnormal function of the brain's blood vessel system.

Vasoconstriction The blood flow in the body that is regulated by the sympathetic division of the nervous system.

Ventral root The motor root of a spinal nerve that attaches the nerve to the anterior part of the spinal cord.

Ventricles A cavity that contains cerebrospinal fluid. There are four ventricles in the brain.

Vertebra (plural, vertebrae) A cylindrical bone that, along with cartilage, makes up the spinal column that encases the spinal cord.

Vertebral arteries In conjunction with the internal carotid arteries, one of two pairs of arteries that supply blood to the brain

Vertebrate Living organisms that have a backbone.

Vesicles A small fluid-containing sac.

Vestibulocochlear nerve The cranial sensory nerve that controls balance in addition to carrying auditory and acoustic information relating to sound from the ear to the brain.

Visceral neuron A type of sensory neuron located in the body's internal organs.

Visceral organs The body's internal organs, such as the heart and lungs, that have nerve fibers and nerve endings that conduct messages to the brain and spinal cord.

Visceral sensations The sensations relating to anything that involves the body's internal organs, such as the glands and the smooth and cardiac muscles.

Vitamins Substances that are fat soluble or water soluble and are essential for the body's normal growth and functions. Vitamins can be obtained through plant and animal foods.

Vitreous humor The fluid that is located in the posterior eye cavity that helps keep the retina in place.

Wernicke's encephalopathy Caused by a severe vitamin deficiency, this brain disease is associated with chronic alcoholism and is characterized by confusion, diminished muscular coordination, and abnormal eye movements.

White matter The nerve tissue located within the central nervous system that contains myelinated axons and interneurons.

Organizations and Web Sites

Alzheimer Research Forum
www.alzforum.org

This Internet-based forum was launched in 1996 in order to develop an online community focused on developing treatments and prevention methods for Alzheimer's disease. This site creates and maintains web-based resources for researchers and facilitates online discussions in order to promote debate and communicate new ideas as a way to contribute to the global effort to cure Alzheimer's disease.

Creutzfeldt-Jakob Disease Foundation
P.O. Box 5312
Akron, Ohio 44334
Phone: (330) 665-5590
Fax: (330) 668-2474
E-mail: crjakob@aol.com
www.cjdfoundation.org/info.html

This foundation was created in 1993 by two families who had each lost a relative to CJD and the neurologist who treated the CJD patients. The foundation seeks to promote the research, education, and awareness of CJD, as well as to reach out to people who have lost loved ones to this illness.

Christopher Reeve Paralysis Foundation
500 Morris Avenue
Springfield, NJ 07081
Phone: (800) 225-0292
www.christopherreeve.org

This organization funds research projects that focus on developing treatments and cures for paralysis caused by spinal cord injury and other central nervous system disorders. This foundation also works to improve the quality of life for people living with disabilities through its grants program, paralysis resource center, and advocacy efforts.

The Johns Hopkins University, Department of Neurology/Neurosurgery
Phipps Building, Room 395
600 N. Wolfe Street
Baltimore, Maryland 21287
Phone: (410) 955-3288
Fax: (410) 955-0207
www.neuro.jhmi.edu

This Web site provides information on various neurological-related disorders from one of America's premier research hospitals.

Michael J. Fox Foundation for Parkinson's Research
Grand Central Station
P.O. Box 4777
New York, NY 10163
www.michaeljfox.org/foundation

This foundation was established in May 2000 by the actor Michael J. Fox following his diagnosis with Parkinson's. The organization's Web site provides information on the disease, as well as research updates on the progress to treat the disease. In addition, the Web site highlights research projects funded by the foundation that focus on further understanding, treating, and possibly offering a cure for Parkinson's.

National Dysautonomia Research Foundation
1407 W. Fourth Street, Suite 160
Red Wing, MN 55066-2108
Phone: (651) 267-0525
Fax: (651) 267-0524
www.ndrf.org

This foundation's Web site provides information about autonomic nervous system-related disorders, as well as updates on research and supported services for patients afflicted with these disorders.

National Headache Foundation
820 N. Orleans, Suite 217
Chicago, IL 60610
Phone: (888) NHF-5552
www.headaches.org

This organization's mission is to educate patients, healthcare providers, and the general public about headaches, as well as promoting research on potential causes and treatments for headaches.

National Institute of Health
9000 Rockville Pike
Bethesda, Maryland 20892
Phone: (301) 496-4000
www.nih.gov

National Institute of Neurological Disorders and Stroke
www.ninds.nih.gov

An agency under the U.S. Department of Health and Human Services, the NIH is the federally funded organization whose mission is to perform medical and behavioral research. The NIH is made up of twenty-seven institutes and centers. The NINDS is one of the NIH's institutes and supports biomedical research on brain and nervous system disorders. Both Web sites contain comprehensive information on nervous system diseases and disorders.

National Organization on Fetal Alcohol Syndrome
900 17th Street NW, Suite 910
Washington, DC 20006
Phone: (202) 785-4585
Fax: (202) 466-6456
E-mail: information@nofas.org
www.nofas.org

This Web site is focused on raising public awareness of fetal alcohol syndrome, as well as providing information on prevention and intervention.

National Spinal Cord Injury Association
Phone: (800) 962-9629
Fax: (301) 881-9817
www.spinalcord.org

Founded in 1948, the NSCIA is dedicated to enhancing the lives of those who have suffered spinal cord injuries. The Web site contains information on various aspects of spinal cord injuries.

National Tay-Sachs and Allied Diseases Foundation
2001 Beacon Street, Suite 204
Brighton, MA 02135
Phone: (800) 906-8723
Fax: (617) 277-0134
E-mail: info@ntsad.org
www.ntsad.org

This Web site contains information on Tay-Sachs and related diseases, in additon to providing support services for individuals and families affected by these diseases. The organization is dedicated to the treatment and prevention of these diseases and supports research and genetic screening, among other initiatives.

Northwestern University, Feinberg School of Medicine, Department of Neurology and Clinical Neurological Sciences
Abbott Hall, 11th floor
710 North Lake Shore Drive
Chicago, Illinois 60611-3078
Phone: (312) 908-8266
Fax: (312) 908-5073
E-mail: neurology@northwestern.edu
www.neuro.nwu.edu

This Web site is filled with basic and comprehensive information on neurological disorders. The department is focused on researching new therapies, while conducting basic and clinical research on causes, treatments, and potential cures of central and peripheral neurologic diseases.

Society for Neuroscience
11 Dupont Circle NW, Suite 500
Washington, DC 20036
Phone: (202) 462-6688
Fax: (202) 462-9740
E-mail: info@sfn.org
www.snf.org

A nonprofit organization, members of the Society for Neuroscience study and exchange research about the brain and nervous system, which includes the study of brain development, sensation and perception, learning and memory, movement, sleep, stress, aging, and neurological and psychiatric disorders. One of the organization's primary goals is to highlight the potential for the study of the brain and nervous system as a distinct field.

Bibliography

Allaby, Michael, ed. *The Oxford Dictionary of Natural History*. Oxford: Oxford University Press, 1985.

Alzheimer Research Forum. www.alzforum.org.

"The Autonomic Nervous System." www.nda.ox.ac.uk./wfsa/html/u05/u05_010 .htm. Nuffield Department of Anaesthetics, University of Oxford, United Kingdom.

"Bell's Palsy." Department of Neurology, Northwestern University Medical School. www.neuro.nwu.edu/meded/cranial/bells.html.

Boshes, Louis D. "Founders of Neurology." Archives at the University of Illinois at Chicago, Department of Neurology, University of Illinois at Chicago. www.uic. edu/depts/mcne/homepage/neurofounders.html.

Carnegie Library of Pittsburgh, Science and Technology Department. *Science and Technology Desk Reference*. Detroit: Gale Research, 1993.

Christopher Reeve Paralysis Foundation. www.christopherreeve.org.

Chudler, Eric. "Neuroscience for Kids." www.faculty.washington.edu/chudler/.

Cleveland Functional Electrical Stimulation Center. http://feswww.fes.cwru.edu/.

Columbia Encyclopedia, 6th ed. New York: Columbia University Press, 2002.

Creutzfeldt-Jakob Disease Foundation. http://cjdfoundation.org/info.html.

Findlen, Paula, and Rebecca Bence. "Early Science Lab." Stanford University, Department of History. www.stanford.edu/class/history13/earlysciencelab/main/index.html.

Green, John. *Human Anatomy in Full Color*. Mineola, NY: Dover Publications, 1996.

Griffiths, Mary. *Introduction to Human Physiology*. New York: Macmillan Publishing Co., 1974.

Hargittai, István. *The Road to Stockholm: Nobel Prizes, Science, and Scientists*. New York: Oxford University Press, 2002.

Holmes, Oliver. *Human Neurophysiology: A Student Text*, 2nd ed. London: Chapman & Hall Medical, 1993.

"How Do Nerve Cells Communicate?" Society for Neuroscience. http://web.sfn.org/Content/Publications/BrainBackgrounders/communication.htm.

"An Introduction to Spinal Cord Injury: Understanding the Changes." Paralyzed Veterans of America. www.pva.org.

Johns Hopkins Medicine, Department of Neurology. www.neuro.jhmi.edu.

Lyons, Albert S., and R. Joseph Petrucelli. *Medicine: An Illustrated History*. New York: Harry N. Abrams Publishers, 1987.

Magner, Lois N. *A History of the Life Sciences*. New York: Marcel Dekker, 1994.

McCourt, Mark. "Historical Origins of Neuropsychology." Department of Psychology, North Dakota State University. www.psychology.psych.ndsu.nodak.edu/ mccourt/website/htdocs/HomePage/Psy486/Historical%20origins%20of%20 neuropsychology/historical_origins_of_neuropsych.htm.

McDowell, Julie L. "More than a Headache." *Modern Drug Discovery*, November 2002, 21–25.

Memmler, Ruth L., Barbara Janson Cohen, and Dena Lin Wood. *Structure and Function of the Human Body,* 5th ed. Philadelphia: J.B. Lippincott, 1992.

Michael J. Fox Foundation for Parkinson's Research. www.michaeljfox.org/ foundation.

NASA Neurolab. "Spotlight on Neuroscience." http://neurolab.jsc.nasa.gov/timeline. htm.

Nathan, Peter. *The Nervous System,* 4th ed. London: Whurr Publishers Ltd., 1997.

National Dysautonomia Research Foundation. www.ndrf.org.

National Headache Foundation. www.headaches.org.

National Institute of Neurological Disorders and Stroke. www.ninds.nih.gov.

National Institutes of Health. www.nih.gov.

National Organization on Fetal Alcohol Syndrome. www.nofas.org.

National Spinal Cord Injury Association. www.spinalcord.org.

National Tay-Sachs and Allied Diseases Association. www.ntsad.org.

Nobel Foundation. www.nobel.se.

Northwestern University Medical School, Neurology Department. www.neuro.nwu. edu.

Porter, Roy, ed. *The Cambridge Illustrated History of Medicine*. Cambridge: Cambridge University Press, 1996.

Sabbatini, Renato M. E. "The Discovery of Bioelectricity: The Field Advances." Center for Biomedical Informatics of the State University of Campinas. www.epub.org.br/cm/n06/historia/bioelectr4_i.htm.

Sanders, Tina, and Valerie C. Scanlon. *Essentials of Anatomy and Physiology,* 3rd ed. Philadelphia: F. A. Davis Company, 1999.

"Scientists Probe Brain and Nervous System Control of Weight, Appetite." Society for Neuroscience. http://web.sfn.org/content/Aboutsfn1/NewsReleases/am 2002_obesity.html.

Stedman, Thomas Lathrop. *Stedman's Medical Dictionary*. Baltimore: Williams & Wilkins, 1995.

Streeten, David H. P. "General Organization of the Autonomic Nervous System." The National Dysautonomia Research Foundation. www.ndrf.org/ans.htm.

Trojanowski, John Q. "Alzheimer's Disease, Parkinson's Disease and Related Brain Disorders: Brief Overview for Patients and Caregivers." October, 1999. University of Pennsylvania, Center for Neurodegenerative Disease Research. www.uphs.upenn.edu/cndr/patients/ADPDoverview.htm.

University of California, Berkeley, Museum of Paleontology. www.ucmp.berkeley. edu/.

WebMD. www.webmd.com.

Willett, Walter C., and Meir J. Stampfer. "Rebuilding the Food Pyramid." December 17, 2002. *Scientific American.* www.scientificamerican.com/article.cfm?articleID=0007C5B6-7152-1DF6-9733809EC588EEDF&catID=2.

Windelspecht, Michael. *Groundbreaking Scientific Experiments, Inventions, and Discoveries of the 17th Century.* Westport, CT: Greenwood Press, 2002.

Wozniak, Robert H. "René Descartes and the Legacy of Mind/Body Dualism." http://serendip.brynmawr.edu/Mind/Descartes.html.

Index

Acetylcholine (ACh), 11–17, 47, 56–57, 94, 96, 108, 139
ADH (antidiuretic hormone), 30
ADP (adenosine diphosphate), 5
Adrenaline, 11
Adrenal medulla, 55
Adrian, Edgar Douglas, 89, 91, 93–94
Afferent nerves, 3
Afferent (sensory) neuron, 5–6
Aging, 56–57, 74, 134–136
Alcohol, 129, 165; effect on nervous system, 136–139; fetal alcohol syndrome, 122
Alzheimer, Alois, 107
Alzheimer's disease, 36, 58, 106–107, 135, 151–152
Amino acids, 11, 30, 39, 126–127
Amphetamines, 144–145
Amyotrophic lateral sclerosis (ALS), 116, 154, 157
Anatomy, history of research, 75–88
Aneurysm, 42; cause of stroke, 112–113
Antibiotics, 93
Aristotle, 77–78, 81
Asbestos, 148–149
ATP (adrensoine triphosphate), 5

Attention deficit hyperactivity disorder (ADHD), 117
Auditory senses, 29–30, 35–37, 40–41, 47–49, 59–74; aging, 134–136; deafness, 71; organs, 69–73; research, 91, 93, 96, 98; Wernicke's area, 35, 37
Autism, 117–118
Autonomic nervous system, 1, 31, 45, 50–52, 94, 96; parasympathetic division, 47, 50–52, 67; sympathetic division, 50–55, 66, 97–98, 135
Axelrod, Julius, 91, 97–98
Axons, 4, 8–10, 16–17, 20, 33, 55–57, 85–86, 145–146; nervous system disorders, 108, 116–117, 158–159, 167

Bárány, Robert, 91, 93
Basal ganglia, 36
Bell, Charles, 85
Bell's Palsy, 85, 111–112
Bergström, Sune K., 92, 100–101
Bernstein, Julius, 86
Bipolar disorder (BD), 118–119, 162–166
Blood alcohol concentration (BAC), 136–138

Blumberg, Baruch S., 91, 99
Body mass index (BMI), 126–127
Borelli, Giovanni, 82
Boyet, Daniel, 91, 95–96
Brain, 27–44; blood supply, 39–42, 50;
    cardiac functions, 47, 50–52; early
    research, 86–87, 93–94, 99; first men-
    tion, 77; hemispheres, 32, 34, 37,
    113, 139; history of research, 75–78,
    86–88; memory, 36, 40–42; motor ac-
    tivity, 34–35; neurotransmitters, 11;
    respiratory functions, 29; scans,
    43–44; sensory functions, 45–58;
    traumatic brain injury, 105–106; ven-
    tricles, 27, 29, 39, 78, 118, 139
Brain stem, 27, 32
Broca, Paul, 87

Caffeine, 136, 140–141
Calories, 125–134
Carbohydrates, 125, 127–128, 130
Carlsson, Arvid, 92, 102
Cell respiration, 5
Cellular fluid, 9, 14–17
Central nervous system, 1, 19, 28, 45,
    59, 151; effect of aging, 74
Cerebellum, 27–29, 43, 71–73, 87,
    118, 138–139, 143, 145, 147
Cerebral aqueducts, 29
Cerebral cortex, 32–34, 36–37, 42–43,
    60, 67, 71, 84, 113, 138–139,
    142–145, 147; Alzheimer's disease
    effects, 107–108
Cerebrospinal fluid, 20, 27, 38–39, 77
Cerebrum, 27, 29, 31, 71; lobes, 29,
    34–37, 61, 71, 95, 118, 139, 162
Choroid plexus, 27
Chronic pain, 117–121
Circadian rhythms, 31–32
Circle of Willis, 40–41, 82
Cocaine, 136, 141, 165
Cohen, Stanley, 92, 101
Computed tomography (CT, also
    known as CAT) scan, 43, 92,
    99–100, 144, 161
Copernicus, Nicolaus, 77

Cormack, Allan M., 92, 99–100
Corpus callosum, 32, 36–37, 100, 139
Cranial nerves, 1, 8, 45–50, 55–56, 59,
    63, 65, 71, 73–74; early research,
    82, 85
Cranium, 27
Creutzfeldt–Jakob Disease (CJD), 123
Cutaneous senses, 35, 61

Dale, Sir Henry Hallett, 91, 94, 96
Deiters, Otto Friedrich Karl, 85–86
Democritus, 76–78, 83
Dendrites, 3, 8–10, 20, 56, 86, 90
Depression, 119, 141–142, 145,
    162–166; Alzheimer's disease, 107
Descartes, René, 80–81, 84
Diabetes, 125–126, 129
DNA, 148, 164; chromosomes, 3
Dopamine, 11–12, 110–111, 115–117,
    138–139, 141–145, 153
Drugs, 136–146, 165
Du Bois-Reymond, Emil Heinrich, 85

Eccles, John Carew, 91, 96
Ecstasy, 136
Efferent nerves, 3
Efferent neuron, 5–6
Endorphins, 99, 110, 142
Epilepsy, 113–115, 118, 162
Erlanger, Joseph, 91, 94–95
Ethology, 91, 98

Facial sensations, 47–49
Fat, 125–134
Fetal alcohol syndrome (FAS), 122
Fluorens, Marie-Jean-Pierre, 43, 86
Food Pyramid, 129–132
Franklin, Benjamin, 83

Gajdusek, Daniel C., 91,99
Galen, 78–79, 81, 83
Gall, Franz Joseph, 43, 86
Galvani, Luigi, 83–84
Gamma aminobutyic acid (GABA), 11,
    12
Gasser, Herbert Spencer, 91, 94–95

Genetic disorders, 69, 110–111, 115–117, 118–119, 123–124, 163–167

Gilman, Alfred, 92, 102

Glia, 27

Glucose, 5, 43, 127, 130

Golgi, Camillio, 90–93

Granit, Ragnar Arthur, 91, 97

Gray matter, 2, 20, 22, 29, 32–33, 36, 38; early research, 86, 91–92

Greengard, Paul, 92, 102

Growth hormone (GH), 30

Growth hormone–releasing hormone (GHRH), 30, 99

Guillain-Barré Syndrome, 108

Guillemin, Roger, 91, 99

Gullstrand, Allvar, 91–93

Gyri, 33

Hartline, Halden Keffer, 91, 97

Headaches, 109, 140–141

Heroin, 142–143

Hess, Walter Rudolph, 91, 95

Hippocampus, 36, 40, 107–108, 118–119, 143–145

Hippocrates, 77–78

Hodgkin, Alan, 91, 96

Hooke, Robert, 2, 81–82

Hormones, 30, 39, 55, 91, 97–99

Hounsfield, Godrey N., 92, 99–100

Hubel, David H., 92, 100

Huxley, Andrew Fielding, 91, 96

Hypothalamus, 27–30, 32, 43, 50–52, 54–55, 64, 91, 95, 99, 110

Interneuron, 5, 7–8

Ions, 14–17, 96, 101, 147–148, 167; discovery of bioelectricity, 83–85

Janssen, Zacharias, 2, 81

Kandel, Eric, 92, 102

Katz, Bernard, 91, 97–98

Lashley, Karl Spencer, 43–44

Leonardo da Vinci, 80

Levi-Montalcini, Rita, 92, 101

Loewi, Otto, 11, 91, 94, 96

Longitudinal fissure, 32–33

Lorenz, Konrad, 91, 98

Magnetic resonance imagery (MRI), 43, 144, 161; autism, 118

Manic depression, 118–119

Marijuana, 136, 143

Medulla, 27, 29, 30, 42, 46, 51, 87, 95; effect of stroke, 113

Meninges, 38, 77; meningitis, 38

Mental disorders, 91, 93, 98, 162–166

Mercury, 147

Microscope, nervous system research, 2, 81–82

Midbrain, 27, 29, 46, 51, 71

Migraines, 108–109

Minerals, 125, 128, 131

Moniz, António Egas, 91, 95

Morphine, 142

Müller, Johannes, 84–85

Multiple sclerosis (MS), 116–117, 121

Myelin sheath, 6–7, 52; nervous system disorders, 108, 116, 117, 159

Neher, Erwin, 92, 101–102

Nerve impulse transmission, 2, 8–17, 19, 25–26, 166–167; early research, 84–86, 94–98, 100–102; muscular, 9, 12–14; respiratory, 9

Nerve tracts, 2

Nervous system disorders, 103–120, 130; aging, 134–136; early research, 77–79, 86–88, 94–95, 98–99, 102

Neuroglia, 6

Neurolemma, 6–7

Neuroscience, history of discovery, 75–102

Neurotransmitters, 5, 10–16, 91, 94, 96, 97–98, 102, 138–145, 147; autonomic nervous system, 56; nervous system disorders, 110–111, 115–118, 120–121, 153, 157, 163–164

Newton, Sir Isaac, 83

Nicotine, 141

Nobel Prize: history, 88–90; neuro-
    science winners, 90–102
Node of Ranvier, 6
Noradrenalin, 11
Norepinephrine, 11–12, 55–56, 97–98,
    115, 138–139, 141, 143–145,
    164–165
Nutrition, 125–134, 163

Obesity, 125–127
Olfactory senses, 35–36, 47–49
Opium, 142–143
Ozone, 149

Paranoia, 95, 145; Alzheimer's dis-
    ease, 107
Parkinson's disease, 36, 110–111, 121,
    152–153, 157; neurotransmitters, 12
Peptides, 4, 11, 12
Peripheral nervous system, 1, 45, 141,
    146; disorders, 108, 121
Phrenology, 43, 86
Pituitary gland, 30, 52, 91, 99
Pons, 27–29, 30, 42, 46, 51; effect of
    stroke, 113
Positron emission tomography (PET)
    scan, 43–44, 144, 167
Postganglionic neurons, 51–53, 55–56
Preganglionic neurons, 51–53, 55–56
Protein, 125–134
Prusiner, Stanley, 92, 102
Purkyne, Jan, 84

Radiation, 147–148
Ramón y Cajal, Santiago, 90–93
Rapid eye movement (REM), 31
Reflex, 19, 25–26, 29, 66; early re-
    search, 81, 93–94; effect of spinal
    cord injury, 103–105; stretch, 20,
    26; visceral, 45–57
Remak, Robert, 84
Renaissance, 79–82
Rodbell, Martin, 92, 102

Sakmann, Bert, 92, 101–102
Samuelsson, Bengt I., 92, 100–101

Schally, Andrew V., 91, 99
Schizophrenia, 95, 98, 162–166
Schwann, Thomas, 84
Schwann cells, 6–7, 84
Seizures, 113–115
Sensory impulses, 32, 34, 35–37,
    59–74; aging, 134–136; drugs,
    136–146; effect of migraines, 109;
    history of discovery, 76–77, 83–85;
    peripheral and autonomic func-
    tions, 45–58
Sensory organs, 59–74; effect of aging,
    74
Serotonin, 11, 109, 115–119, 141, 143,
    145, 146, 164–165
Sherrington, Sir Charles Scott, 89, 91,
    93–94
Skull, 27, 38
Sleep, 31
Smelling senses, 61–63
Smith, Edwin, 77
Speech, 35–37; Broca's area, 35, 37;
    early research, 87–88; effect of
    stroke, 42, 113
Sperry, Roger W., 92, 100
Spinal cord, 19–26; dorsal root,
    20–22; ventral root, 20–23
Spinal cord injuries, 7, 103–105, 121;
    early research, 78–79, 151; future
    research, 155–162
Spinal nerves, 1, 20–24, 45, 50–52,
    59; early research, 85; nerve
    plexus, 21–11, 24
Spurzheim, Johann 43
Stem cells, 151–155, 159; Parkinson's
    disease, 153
Steno, Nicolaus, 82
Stroke, 34, 40, 42, 112–113
Sulci, 33
Synapse, 8–10, 12–14, 52, 55–56, 91,
    94, 96, 118, 140, 144, 158; thresh-
    old level, 9

Taste senses, 35–36, 47–49, 59–74;
    aging, 134–136
Tay-Sachs disease (TSD), 110

Thalamus, 27, 32, 52, 95, 120–121
Thoracic nerve, 48–49
Tinbergen, Nikolaas, 91, 98
Tourette, Georges, 115
Tourette Syndrome (TS), 115–116
Traumatic brain injury (TBI), 105–106

Van Leeuwenhock, Antony, 2, 81
Van Waldeyer, Wilhelm, 86
Vane, John R., 92, 100–101
Vertebrae, 19–20, 103–104, 155, 160
Vesalius, Andreas, 79
Visceral organs, 45, 47, 52–54; sensations, 63–64
Visual senses, 35–37, 40–41, 47, 49, 64–69; aging, 134–136; color blind-

ness, 69; organs, 64–69; research, 91–93, 97–98, 100
Vitamins, 125, 128, 131
Von Békésy, Georg, 91, 96
Von Euler, Ulf, 91, 97–98
Von Frisch, Karl, 91, 98
Von Helmholtz, Hermann, 84

Wagner-Jauregg, Julius, 91, 93
Wald, George, 91, 97
Waller, Augustus, 85
Wernicke, Carl, 87–88
White matter, 2, 20, 22, 33, 36, 38; early research, 86; sensory tracts, 59
Wiesel, Torsten N., 92, 100

About the Author

JULIE McDOWELL is an independent scholar and science journalist. She is coauthor of *The Lymphatic System* in Greenwood's *Human Body Systems* series. She is also a former assistant editor for two science publications, *Today's Chemist at Work* and *Modern Drug Discovery*.